Number 1 · 1982 · Volume 2

Acta Biotechnologica

Journal of microbial, biochemical and bioanalogous technology

Akademie-Verlag
Berlin

ISSN 0138-4988
Acta Biotechnol., Berlin 2 (1982) 1, 1—116
EVP 30,— M 31007

Instructions to Authors

1. Only original papers that have not previously been published will be accepted.
   Manuscripts may be submitted in English, German or Russian but always in Duplicate. The name of the institute (inclusive full address) from which the manuscript orginates should be stated below the author's name(s). The authors are responsible for the contents of their contributions.

2. Original papers should not consist of more than 20 typewritten pages (2 lined-spacing), including references, tables and figures; short original communications may have a maximum of 6 typewritten pages.

3. Each paper should be preceded by two summaries, in each case one in english.

4. Latin names of species as well as passages to be printed in italics for greater emphasis should be signed with weved line. Please use only units and symbols of the SI-system.

5. Tables may be used to effect a reduction of the text or to make assimilation easier. They should be numbered consecutively throughout and supplied with a brief heading. They shouldn't be written among the text but on separate sheets.

6. Number and size of illustrations should be limited to the essential minimum, they should be numbered consecutively and brought on separate sheets.
   Line drawings including graphs and diagrams should be drawn in black ink. Half-tone illustrations should be supplied as white glossy prints. Figure legends are to be typed in sequence on a separate sheet.
   Authors name is to be written on to the rear of each sheet.

7. References, listed at the end of the contribution, should contain only works cited in the text. They should be numbered in the order in which they are at first mentioned in the text. Please give surnames and initials of all authors, name of journal abbreviated according to "Chemical Abstracts — List of Periodicals", volume number, year of publication, issue number or month, first page number; books are to be cited with full title, edition, volume number, page number, place of publication, publisher and year of publication.

8. Notes on the text may be given as footnotes on the relevant page.

9. 50 reprints are free of charge.
   Additional reprints may be ordered.

10. The author receives 2 galley proofs for correction.
    They are send back to the redaction
    (Dr. Dimter DDR-7050 Leipzig, Permoserstr. 15)
    as soon as possible.

Acta Biotechnologica

Journal of microbial, biochemical and bioanalogous technology

Edited at
Institute of Technical Chemistry of the Academy of Sciences of the G.D.R.; Leipzig
and
Institute of Technical Microbiology; Berlin
by
M. Ringpfeil, Leipzig and G. Vetterlein, Berlin

Number 1

1982

Volume 2

AKADEMIE-VERLAG · BERLIN

"Acta Biotechnologica" publishes reviews, original papers, short communications and reports out of the whole area of biotechnology.
The journal shall promote the foundation of biotechnology as a new, homogeneous scientific field. According to biotechnology are microbial technology, biochemical technology and technology of synthesyzing and applying of bioanalogous reaction systems.
The technological character of the journal is guarenteed thereby that microbial, biochemical, chemical and physical contributions must show definitely the technological relation.

Terms of subscription for the journal "Acta Biotechnologica"
Orders can be sent

- **in the GDR:** to Postzeitungsvertrieb or to Akademie-Verlag, DDR-1086 Berlin, Leipziger Straße 3–4;
- **in the other socialist countries:** to a book-shop for foreign languages literature or to the competent news-distributing agency;
- **in the FRG and Berlin (West):** to a book-shop or to the wholesale distributing agency Kunst und Wissen, Erich Bieber OHG, D-7000 Stuttgart 1, Wilhelmstr. 4–6;
- **in the other Western European countries:** to Kunst und Wissen, Erich Bieber GmbH, CH-8008 Zürich, Dufourstraße 51;
- **in other countries:** to the international book- and journal-selling trade, to Buchexport, Volkseigener Außenhandelsbetrieb der DDR, DDR-7010 Leipzig, Postfach 160, or to Akademie-Verlag, DDR-1086 Berlin, Leipziger Straße 3–4.

Acta Biotechnologica

Herausgeber: Institut für technische Chemie der AdW,
DDR-7050 Leipzig, Permoserstr. 15
(Direktor: Prof. Dr. Manfred Ringpfeil) und
Institut für Technische Mikrobiologie
DDR-1017 Berlin; Alt-Stralau 62
(Direktor: Dipl. Ing. G. Vetterlein).

Verlag: Akademie-Verlag, DDR-1086 Berlin, Leipziger Straße 3–4;
Fernruf: 2236221 und 2236229; Telex-Nr.: 114420; Bank: Staatsbank der DDR, Berlin,
Konto-Nr.: 6836-26-20712.

Redaktion: Dr. Lothar Dimter (Chefredakteur),
Martina Bechstedt (Redakteur), DDR-7050 Leipzig,
Permoserstr. 15; Tel.: 6861255.

Veröffentlicht unter der Lizenznummer 1671 des Presseamtes beim Vorsitzenden des Ministerrates der Deutschen Demokratischen Republik.

Gesamtherstellung: VEB Druckhaus „Maxim Gorki", DDR-7400 Altenburg.

Erscheinungsweise: Die Zeitschrift „Acta Biotechnologica" erscheint jährlich in einem Band mit 4 Heften. Bezugspreis eines Bandes 120,– M zuzüglich Versandspesen; Preis je Heft 30,– M. Bestellnummer dieses Heftes: 1094/2/1.

AN (EDV) 42133

Acta Biotechnologica 2 (1982) 1, 3—41

Status and Developments of Animal Cell Technology using Suspension Culture Techniques

H. W. D. KATINGER[1] and W. SCHEIRER[2]

1 University of Agriculture, Institute for Applied Microbiology,
A 1190 Vienna, Peter Jordan Straße 82, Austria
2 Sandoz Research Institute, Department for Immunobiology,
A 1235 Vienna, Brunner Straße 59, Austria

Summary

In this review the state of the art in animal cell technology, using suspension culture techniques is updated as far as the end of 1980. We have tried to discuss, on a broad basis, the current status and potential developements of both, the purely biological and the biochemical engineering aspects which may be important to improve the performance and design of animal cell technologies. The process economics could be considerably improved by the use of transformed animal cell substrates, the use of cheaper cultivation media and by methodological and engineering means. Most of these aspects are in the state of realization. Nevertheless, for a great variety of biological active substances which do not require co — or posttranslational processing, recombinant DNA-techniques (e.g. genetic engineering) are a promising alternative for the production of animal cell derived substances.

Zusammenfassung

Mit diesem Beitrag wird versucht, eine Übersicht über den Entwicklungsstand der „Technologie mit tierischen Zellen“ (begrenzt auf Verfahren mit Suspensionskulturen) mit Status Endc 1980 zu vermitteln. Wir haben dabei versucht, zur Zeit schon etablierte Verfahren wie auch potentielle Entwicklungsmöglichkeiten dieser Technologie aufzuzeigen und zu diskutieren. Es wurden hierbei sowohl die rein biologischen Aspekte sowie die sich daraus ergebenden Konsequenzen für die biologische Verfahrenstechnik im Hinblick auf die Verfahrensabwicklung und Möglichkeiten der Prozeßverbesserung analysiert.
Die Technologie mit tierischen Zellen kann durch den vermehrten Einsatz transformierter Zellen als Produktionsvehikel, die Entwicklung billiger Nährmedien und die Verbesserung der Massenkultivierung auf methodischer wie verfahrenstechnischer Ebene noch wesentlich verbessert werden. Viele der genannten Verbesserungsmöglichkeiten sind im Entwicklungsstadium. Für die Produktion einer Reihe von bioaktiven Wirkstoffen aus tierischen Zellen, welche für ihre Biosynthese co- oder posttranslationale Modifikationen nicht erfordern, sehen wir die bessere Alternative in der DNA-Technologie (z. B. mittels der Gentechnik).

Preface

This review article is the full text of an invited lecture held at the "Chemap Cell Culture Symposium", sponsored and organized by the Chemap AG in January 1980. The state of the art in animal cell technology, using suspension culture techniques has been updated as far as the end of 1980. It is mainly addressed to biotechnologists who are „non-specialists“ in the field of animal cell technology.

Table 1. Potentially useful products derived from animal cells at present

Product	Cells used	Susp.	Monol	Micro carr.	Product scale	Reference
VACCINES						
Foot and Mouth Dis.	BHK 21	×	×		Industrial	66
Foot and Mouth Dis.	BHK 21			×	Pilot	73
Polio	Primary Kidney		×	×	Industrial	129
Polio	Hum. Dipl. Fibrobl.		×		Industrial	45
Rubella	Hum. Dipl. Fibrobl.		×		Industrial	45
Measles	Hum. Dipl. Fibrobl.		×		Industrial	45
Rabies	Hum. Dipl. Fibrobl.		×		Industrial	45
Tick-Borne-Encephal-itis	Hum. Dipl. Fibrobl.		×		Industrial	45
Fish Viruses	Fish Cells	×	×		Pilot	59
Epstein-Barr-Viruses	Lymphoblasts	×			Pilot	60
Cytomegalovirus	Wi 38		×		Pilot	97
Insect Viruses	Insect Cells		×		Laboratory	75
HORMONES						
ACTH	Pitutary tumor	×	×		Laboratory	115
Prolactin	Pitutary lines	×	×		Laboratory	115
Growth hormone		×	×		Laboratory	115
Gonadotropin	Trophoblastic line				Laboratory	40
Anterior Pitutary Horm.						64
Prostaglandins						37
Parathormone	Parathyroid Cells	×			Laboratory	80
ENZYMES						
Collagenase	Skin fibroblast Synovial Cells		×		Laboratory	10
Collagenase and Plas-minogen Activator	Synovial Cells		×		Laboratory	127. 132
Diff. Enzymes						100
INTERFERON	Lymphoblasts	×			Industrial	47. 61
INTERFERON	Fibroblasts		×	×	Industrial	23
Lymphokine	Lymphoblasts	×			Pilot	92
Surface Antigens	Lymphoblasts	×			Pilot	2. 26. 21. 114
Leukemia Cells for Therapy	Lymphoblasts	×			Pilot	104
Specific Antibodies	Lymphoblasts	×			Laboratory	113. 123. 135
HYBRIDOMAS	Lymphoblasts	×			Laboratory	74 (Review)
HLA-Antibodies	Lymphoblasts	×			Laboratory	15. 25
Tumor-Antigen-Antibody	Lymphoblasts	×			Laboratory	1. 56. 103
Neutralizing Virus Antibody	Lymphoblasts	×			Laboratory	109
Enzymes	Lymphoblasts	×				131
IGE	Lymphoblasts	×			Laboratory	12

1. Introduction

Animal cells have been routinely cultivated in small bench scale cultivation units for a long period of time as research tools for the study of general biological principles, for cancer research, other medical and diagnostic applications and as substrate for virus multiplication. They have also been successfully used as highly sensitive test systems for the evaluation of various beneficial or hazardous substances. Cultured animal cells are often very sensitive in their response to even minute environmental stimuli; for example a few picomoles of a toxin per millilitre of cultivation broth is sufficient to induce drastic changes in cell morphology [62, 67]. Such unexpected manifestation of the sensitivity of animal cell cultures, whatever their benefit for a biological test system, is a major cause for the dogmatism prevalent in the business of animal cell technology and the exploitation of its potential usefulness for society.

Advances in basic research with animal cells confront us with a steadily increasing palette of biologicals of potential (therapeutic and diagnostic) use. In reality, however, only a humble fraction of the vast array of possibilities (e.g. production of viral vaccines) has so far come to commercial realization. The production (in sufficient quantities) of the majority of the biologicals referred to in Table 1, would be desirable if economically feasible.

The dramatically increasing public demand for these new biologicals, together with the rather timid progress in the search for a technologically satisfying solution to the supply problem has revealed a gap in our knowledge which calls for a critical evaluation of the *status quo*. Thus, this contribution will consider the state of the art in large scale suspension cultivation techniques of animal cells, together with its relevance to the production of the useful biologicals. Special attention is paid to the major demands of mass cultivation of animal cells and (*a*) the nature, quality, maintenance and control of animal cell substrates; (*b*) the nature of growth media and associated problems; (*c*) the cultivation techniques, equipment and methods applied for mass propagation. The rationale behind solving these problems will be strictly biotechnological.

2. General Background

Before attempting to consider in detail all the scientific and practical aspects involved in such a complex matter as animal cell technology, a few definitions and simplifications are necessary. The aim of biotechnology, briefly, is to interrelate economical, biological, biochemical and engineering techniques as elements of a novel art to produce scientifically relevant and commercially valuable results. It is the economic or commercial element inherent in this definition which complicates the whole issue and is responsible for the difference between mere process development and the development of a technological process.

Fig. 1 shows a simplified flow sheet of technological design and its interrelationships. It may be used as a guide in following up the main routes of process development work and the resultant feedback to decision making. The complicated machinery of establishing new, or improving existing technologies, is subject to distinct hierarchical controls. The "go ahead" has to come from the economic environment where the market for a product, its socio-economic benefits, and the demand-cost-price-profit-relations all have to be critically evaluated. Once the feasibility has been assessed for a given product, the ball moves back into the scientific environment for further evaluation. It passes through several stages of development (II to IV in Fig. 1) where basic insights — the "know why" — and finally the "know how" of the process are acquired. The information obtained from each stage of the process development act has some kind of negative or positive feedback influencing further advances. The resulting decision must fall within the existing regulations concerning the product registration which should undergo permanent reexamination in the light of recent advances. This process would be essential for establishing new technological methods.

A central feature of any biological process is the production strain since it accounts for both (*a*) the final process result because it is the vehicle by which the production can be maximized and (*b*) the total process expenditure, because it is either satisfied with a simple environment or it demands a very nutritious and high complex chemical and physical environment, which is necessarily expensive. Strain selection and improvement is, therefore, of prime importance in biotechnology (step II in Fig. 1).

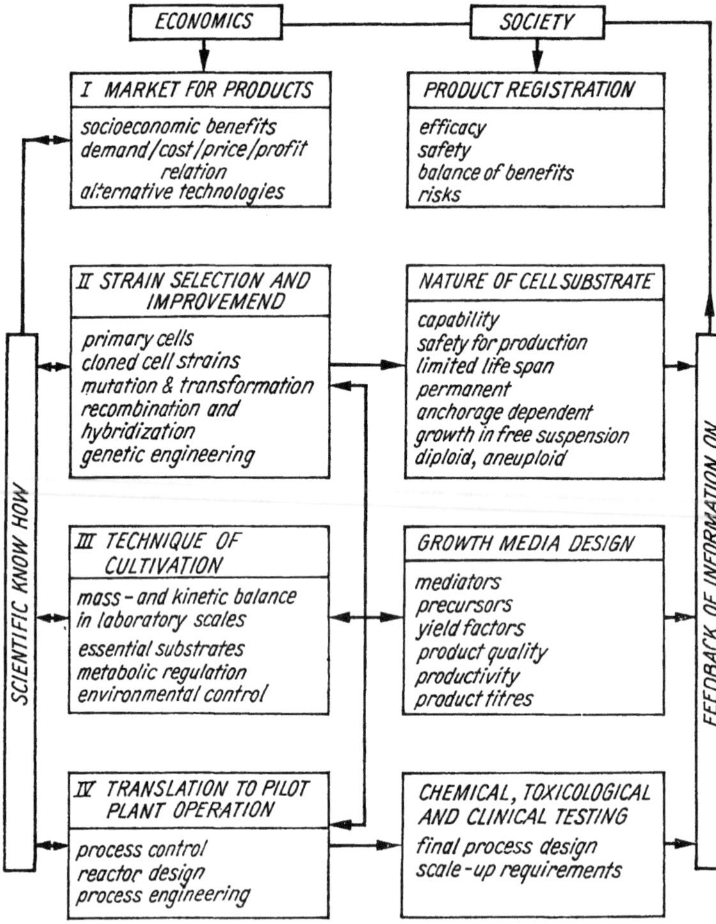

Fig. 1. Interrelation between techniques, economics, society and cell culture products

In microbial technology use has been made of all the techniques available from genetic science (compare Tab. 2). Tactics in production strain improvement programmes have not been exclusively restricted to the selection of natural variants (as a consequence of biological evolution) or cloning of these variants in order to guarantee their genetic homogeneity. Stress has also been laid on the creation of artificial biological variants, either by means of mutation — and/or by the use of transformation procedures, by sexual or parasexual (somatic) recombination techniques (for gene-exchange and genamplification)

or by recombinant DNA-techniques (by gene-manipulation). Such intentionally transformed microbes are currently used as vehicles to produce food or pharmaceuticals. In animal cell technology, on the other hand, the general situation is more complicated and the philosophy behind their manipulation is quite different.
Most of the basic techniques applied to the manipulation of microorganisms including genamplification and gene-exchange by somatic cell recombination [74a] and even by the manipulation and transfer of genes into animal cells by genetic engineering techniques [16a, 82a], are applicable to animal cells. However, the status quo in animal cell technology is different. Until now only primary cultures of cells freshly explanted from normal tissues, and a few diploid cell strains with finite life spans, have been approved by the public regulatory authorities for use as cell substrates in the production of viral vaccines or other biologicals destined for human therapeutic use [93]. Continuous line cells (transformed) or any other type of non-normal cell substrates are not currently accepted.

Table 2. Tactics in process optimization (detailed structure omitted)

Approach	Goals	Strategy	Levels of Operation
Strain Improvement	Improvement & Maintenance of Genetic Potentials	Selection Cloning Mutation & Transformation Recombination (Sexual, Somatic) Recombinant DNA techn. (genetic engineering)	Evolutionary Variants Homogeneity Genetic Diversification & Deregulation Gene-Exchange Genamplification Manipulation & Transfer of Genes
Environmental Control and Interaction by Nutritional, Cultural and Technical Means	Utilization of Genetic Potentials (Quality & Yield of cell Mass and Products)	Growth Media Induction Cell Cycle Control Method & Techniques of Cultivation Interactive Control of Parameter via Closed Loop and/or Gate Way Systems	Interaction on Transcription Translation & Post-translation Levels

In earlier years, when the current status of genetic science purification methods and safety testing of products were not available, these restrictions were necessary. Today this "safety ideology" inhibits the development of animal cell technology and restricts the application of a potentially promising beginning. It is pertinent to the discussion to review some of the many arguments against this point of view.

(1) Some very useful biologicals, such as monoclonal antibodies, are exclusively obtainable from non-normal line cells (hybridomas see Table 1); their possible therapeutic human exploitation [82b] can therefore not be realized under the existing legislation.

(2) More than 30 years of microbial production strain improvement have taught us that real progress in the mass production of primary or secondary metabolites is only possible when diverse methods of increasing the production are applied in addi-

tion to the selection of naturally occurring bacterial substrates. As long as the use of these tools of manipulation for the improvement of animal cell substrates is prohibited, the most powerful instrument of process development remains useless.

(3) Without reexamination of the current restrictions surrounding the acceptability of tissue culture systems for the production of biologicals, animal cell technology will not overcome a developmental stage comparable to the early days of antibiotic production. The exclusive use of primary cell cultures or cell strains with finite life spans as production cell substrates, with all their intrinsic disadvantages, are totally antagonistic to the development of economic mass production systems (compare Chapter 4.1). Problems connected with the supply of fresh cell tissues and the obvious risk of contamination from such a supply [27, 112] have resulted in complicated control methods, extremely costly culture media and the realization of inadequacies in existing techniques of cultivation (compare Chapter 4.2) all of which are largely consequences of the current regulations.

It is simply inevitable that a new area of animal cell technology demanding the application of permanent and genetically manipulated cell substrates will be created. The development of such new technologies will be stopped in the early stages unless clear guidelines are introduced for the currently unanswered questions concerning the acceptability of such cell substrates.

The potential benefits of new cell culture products such as interferon from lymphoblastoid cells have renewed the interest of the scientific community in giving permission for the use of permanently growing cell substrates. In the autumn of 1978 this complex issue was the subject of a specialized Symposium [94]. The unanimous opinion expressed at this meeting was, that the use of transformed cell substrates has obvious practical advantages and only theoretical disadvantages for the production of biologicals in general. This issue is particularly true for products such as killed viral vaccines and viral extracts as well as for all proteins which can be purified sufficiently. In this context, a quotation from the report of the *ad hoc* Committee on Karyological Controls states that "... the potential risks of using karyologically abnormal cells are related to the possible contamination of the product with cellular nucleic acid. Control authorities should, therefore, consider the purification procedures used in the manufacturing process and the confidence with which one can exclude the presence of cellular nucleic acid in the final product...".

Scientists, as well as producers of such products, should learn from this recent reevaluation and be encouraged to develop their processes with methods of strain improvement similar to those which microbial technology has successfully applied in the past. The consequence of such development would lead to more numerous technologies using cell substrates rendering high yields on less expensive culture media.

Many scientists attempting the production of biologicals of animal origin hold the application of recombinant DNA techniques (e.g. genetic engineering) superior to the technology of animal cell cultivation. By these techniques the microorganisms can be constructed by transformation with manipulated genes in such a way that they are able to produce substances previously foreign to them.

These recombinant DNA techniques have already been applied successfully to the production of such important human pharmaceuticals as human insulin [18], somatostatin [43], human interferon [83] and urokinase, all using *Escherichia coli* as host micro-organism.

It is reasonable to assume that these promising initial results represent merely the first landmarks on the way to a new biotechnological era. One conclusion stemming from this work may be that recombinant DNA techniques in microorganisms could solve the various problems related to animal cell mass cultivation at the microbial level and there may, therefore, be no future in animal cell technology.

Although many authorities may support the latter opinion, the central issue of whether recombinant DNA-techniques in microbes or technologies using animal cells are to be favoured deserves critical consideration. However, most of the arguments both for and against are speculative in nature. In this context several main questions have to be considered.

a) To which extend is it possible to increase the product titres, either in genetically engineered microorganisms or in animal cells?

b) Do the production strains, the recipiants for foreign structural genes, have the cellular structural capacity necessary for co- or posttranslational processing of the desired biologicals such as glycoproteins? Prokaryotic microorganisms, for example, are known not to possess the necessary physiology for such biochemical functions, whereas, it has not yet been conclusively shown whether eukaryotes, such as yeasts, do. However, some products may not require co — or post-translational modification. For example, glycolysation is not required for interferon production.

c) Is the product secreted or has it to be extracted from the biomass? In *E. coli* and in other gram-negative bacteria, for example, desirable products are often excreted along with endotoxins.

d) Is mass cultivation possible under reasonable conditions?

Other parameters, particularly the generation times (for both bacteria and animal cells), are, in most cases, of only minor importance when compared to the overall cost of the total process.

Before we can make a comparable analysis between microbial and animal cell technology, the latter must first be given more freedom to display its virtues. One well-established fact supports the use of genetically manipulated microorganisms — their mass cultivation is easier. Theoretically speaking, it should be simpler to transduct the genetic information necessary for the expression of one particular product into a relatively easily cultured microorganism rather than to achieve comparable production by modifying an animal cell with high nutritional demands into an uncomplicated production strain.

3. Products from cells grown in suspension culture

From a superficial glance over the available data, it is apparent that, of the great multitude of animal cell culture products, fully established technologies exist only for viral vaccines. Most of these (i.e. those for human use) are derived from anchorage dependent cells, (Table 1) although their production would be, strictly speaking, possibly cheaper with cell substrates grown in free suspension. Two important factors govern their use: (*a*) A licence (for human use) for cell line-derived products cannot be obtained for those countries in which legislation presently exists. (*b*) The strategy of applying live viral vaccines has been favoured for a long time and has thus so far excluded the use of transformed cell-line products *a priori*.

The situation with other animal cell biologicals (as shown in Table 1), which have not yet passed an — optimistic — "preproduction" stage, is very similar. The fear of transformed cell product registration being unacceptable to the public has created a vicious circle in which those commercially interested parties proceed only hesitantly and cautiously within the technologically problematic area of anchorage-dependent cell substrates.

The production of vaccines against foot and mouth disease in cattle (FMD) has, therefore, remained the most cited technology. It uses transformed cell substrates (baby hamster kidney) grown in free suspension. This technology has acquired immense economic importance and is being carried out in bioreactors of several thousand litres capacity [29, 82].

One could doubt, that, with products for human use, there is going to be little, if any, chance to develop technologies based on animal-cell suspension culture technologies. However, the situation is not as bad as it may appear at first sight, as the current trend appears to favour the use of suspension culture techniques. The liberal attitude with respect to the use for heteroploid cell substrates for the production of biologicals, which originated at the Lake Placed Conference [94], could lead to the advanced use of cell line substrates and of suspension culture techniques. New groups of products in the early stages of being commercialised (compare Table 1) are most likely to become candidates for mass production in suspension cultures (or alternatively for recombinant DNA-technology). The potential role of interferon requires no further elaboration. The ecological potential of insect viruses for biological pest control has also been investigated [75] and is of great importance if we keep in mind that "biological agriculture" is going to be in ever-increasing demand by our modern society. The essential prerequisite for all of these productions is an inexpensive mass cultivation in large vessels.

All of the chemically well defined biologicals compiled in Table 1 (hormones, interferon or other lymphokines, urokinase and possibly viruses), which require mass cultivation, are also candidates for recombinant DNA-technology. The production of highly specific monoclonal antibodies, or similarly complicated structures, is however, for the forseeable future, exclusively restricted to animal cell substrates. They are produced by hybridomas or transformed lymphoid cells and are of special concern to society because of their wide range of uses. The basically simple technique of hybridization, originally developed by LITTLEFIELD [65], and further developed by KÖHLER & MILSTEIN [61a], as well as the highly specific reagents produced by these hybridomas call for technological exploitation. They enable us to produce highly valuable analytical tools for the assay of cell substances which are immunogenic. Other applications of hybridoma products also have been proposed which could be produced without licencing problems for use as diagnostic tools in the identification of drugs, hormones, proteins, viruses and bacteria, tissue typing [82b] blood grouping [57, 101], as well as for obvious therapeutic uses. The application of highly specific antibodies as ligands for affinity chromatography in order to solve sophisticated industrial scale separation problems has suggested potentially important benefits of this biotechnology which, as yet, cannot be fully evaluated [108]. Work is in progress on all of these aspects and, in most cases, results are promising.

The establishment of human-human hybridomas producing homologous monoclonal antibodies has been recently reported [91a]. The application of monoclonal antibodies at the clinical level, which has been recently reported for heterologous antibodies [82b], suggests, that similar applications, using homologeous proteins, may soon be possible. *In vitro* cultivation of these hybridomas on "large scale" is of great importance in this context, since the current practice of cultivation in stationary suspension yields relatively low cell densities and a limited recovery of antibodies. On the other hand, the *in vivo* cultivation in mice (as ascites tumor), which is currently the most often applied method for the production of monoclonal antibodies, is also unreliable for the production of human therapeutics as well as for any form of mass production. However preliminary results in our own, and in other laboratories [21a], indicate that the continuous large scale cultivation of hybridomas as free suspension cultures (in combination with continuous dialysis of the reactor contents) should be possible with reasonable yields of the secreted antibodies (between 100 and 400 mg/l).

Presently, it is without doubt dangerous to make cost calculations and extrapolations based on such a developing field and such data should be considered with scepticism. Nevertheless, crude estimations based on the antibody titres currently obtained by fermentation show that the product unit costs of both fermenter-derived and ascites-derived antibodies are comparable. However, extrapolations to large scale production suggest that the permanent keeping, inoculation and harvesting of more than 10.000 mice could be replaced by one continuously operated 1000 l fermentor.

4. Sources of cell substrates, their routine management and related consequences for technological use

The general expenditure for cell cultivation depends by far on the type of cell to be grown. The technique of cultivation to be applied (whether the cells need attachment to a solid supporting matrix or are able to grow in free suspension), the right choice of cultivation medium and the amount of work to be invested for their routine management and control, varies greatly with the cell type cultured. An exact classification of cultured cells is difficult, and, therefore, it will be helpful to begin with a brief outline of the type of cells currently in use.

4.1. Sources of cell substrates

All cultured cells were originally derived from animals or by mechanical and/or enzymatic dissociation of explanted healthy or malignant tissues or by *in vivo* perfusion (compare Fig. 2). The tissue fragments are placed into a suitable medium and allowed to develop into primary cultures of cells with either "normal" (diploid) or "non-normal" (transformed or tumor-derived) karyotypes. Primary cultures show a variety of different behaviours depending on the type of tissues and the environmental conditions. Some cultures after passaging die within a few days, whilst others persist in culture without showing any morphological or biochemical deviations. Still others begin to divide rapidly and continue this growth for some time before they die after a period which varies from a few weeks to several months. Cells of embryonic origin usually remain alive for the longest period of time. Some diploid cell strains tend to retain their diploid status for up to 80 or even more divisions [46, 98]. Various types of primary cells along with a few characterized serially propagated human diploid cell strains have been licensed for use in producing human viral vaccines and other biologicals [45]. All of these cells belong to the anchorage dependent types with the exception of cells derived from blood.

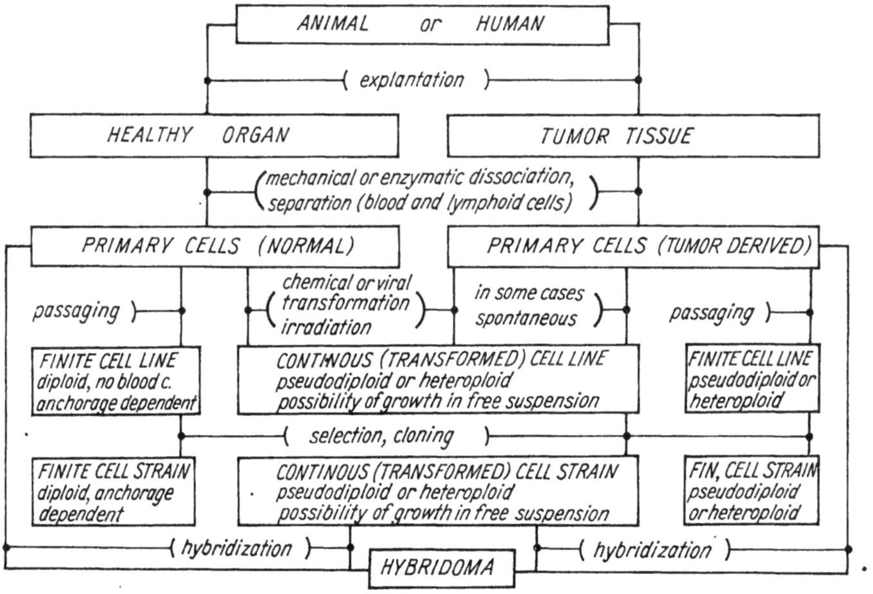

Fig. 2. Nomenclature and derivation of different kind of cells

Table 3. Basic criteria for the design of cell culture system (minor details omitted)

CRITERIA	Variability Range of all Culture Systems
product and kinetics of product formation	CELL MASS CELL PRODUCTS permanent — induction necessary — cell cycle related
desired scale of operation	research scale → industrial scale
cellular properties	limited life span → unlimited life span anchorage dependent cell types → suspension cell type
biochemical engineering tecniques	MATRIX CULTURE → FREE SUSPENSION CULTURE (improved reactor performance and improved environmental control) MATRIX STATIONARY, MEDIUM STATIONARY monolayers, multilayers in bottles etc. ↓ MATRIX STATIONARY, MEDIUM MOVED packed columns etc. (spheres, fibres, spirals etc.) ↓ MATRIX MOVED, MEDIUM MOVED roller bottles, rotating discs or tubes etc. ↓ MATRIX SUSPENDED, MEDIUM AGITATED microcarrier culture ↓ PSEUDOMATRIX suspension of flocs, cell aggregates ↓ MATRIX NOT NECESSARY FREE SUSPENSION CULTURE cells stationary — cells suspended
cultivation methods available	CLOSED SYSTEMS → OPEN SYSTEMS (more economical utilization and production efficiencies) cells and medium retained BATCH CULTURE ↓ cells retained, medium renewed partial: FED BATCH CULTURE, DIALYSIS CULTURE — total: PERFUSION CULTURE ↓ cells and medium renewed SEMICONT. CULTURE — CONTINTINUOUS CULTURE chemostat, turbidostat

The fate of serially passaged and propagated cells, subsequent to primary culture, varies enormously and this results in a continuous spectrum with the fully "normal cell" and the fully "non-normal cell" at the extremes (compare Fig. 2). The term "normal" is used here to describe cells displaying major characteristics that do not differ from cells which occur in healthy animals (compare [79, 105] for definition).
Most of the cultured cells suitable for free suspension culture belong to the "non-normal" transformed type of cell lines. In contrast to normal cells, they sometimes display similar growth patterns to microorganisms i.e. grow in free suspension, have unlimited life span and sometimes propagate in less complex growth media than that required by normal cells. Line cells established directly from tumours (i.e. in vivo "transformed") frequently retain specialized functions (for example: overproduction of enzymes) during prolonged subcultivation for longer periods of time than do those cells which have been intentionally transformed *in vitro* by chemical, physical, viral or other means.

4.2. Cultivation and management of cells

A wide range of different cell culture systems is used for cell propagation and the environmental control of cellular functions *in vitro*. The major aspects in the design of potentially possible *in vitro* systems, and the variability of design criteria from a biochemical engineering (i.e. cultivation techniques) as well as from a methodological (i.e. cultivation methods) standpoint are briefly summarised in Tab. 3. The whole issue is very complex. Criteria such as whether the goal of the cultivation is simply to achieve a certain amount of cell mass, or cell mass in an optimal physiological state of development or is to obtain a cell-derived product with defined kinetics of product formation (e.g. to growth-related, induction-dependent and/or specific cell cycle-related) is decisive in the choice of a proper cultivation method. Cellular properties such as whether cells belong to the anchorage dependent cell types and thus require a growth supporting matrix, or whether they can be grown in free suspension, will greatly affect the process design at the biochemical engineering level (i.e. applying the most suitable cultivation techniques). Other cellular properties, such as whether the cells have a limited life span or not, will greatly affect the amount of work necessary for their routine maintenance (compare 4.2.1. and Figure 3). The situation is further complicated by the fact that a distinct classification and quantification of these design criteria mentioned above is virtually impossible since the complex range of process requirements is interrelated with that of the cells. Therefore "real" cell culture systems used in animal cell technology combine, more or less strictly, the "ideal" principles shown in Tab. 3. To date, a wide variety of technical realizations of these principles has been published or patented. However in this article, only suspension culture and related techniques are considerd. From the following sections it will be seen that the more animal cells tend to develop themselves away from the "normal state" (i.e. diploidy, finite life span, anchorage dependence etc.), the more suitable the will become as vehicles for technological use.

4.2.1. Maintenance of stock cultures and inocula

It is critical for a technologically relevant production to be provided with cells in sufficient quantity with an assurance of maintained physiological properties. This requirement is easily met with permanent cell lines. In order to store the cell material, once its importance in biological production has been recognized, the use of liquid nitrogen is usually most satisfactory.
Maintenance of cells with a limited life span require considerably more precautions in order that a technology established with relatively high expenditures does not run out of cells. The simplest scheme for the maintenance of liquid nitrogen stocks of cultures of

these types of cells, which guarantees an almost inexhaustive supply of stable cell material with the minimum requirement of labour, is shown in Fig. 3.
The limit of storage time for cells in liquid nitrogen is not yet known [3]. It is very important to start this scheme of maintenance of stock culture with cells passaged as early in the selection programme as possible. For additional safety requirements it is recommended to store a few of the stock ampoules containing very valuable cell material in a

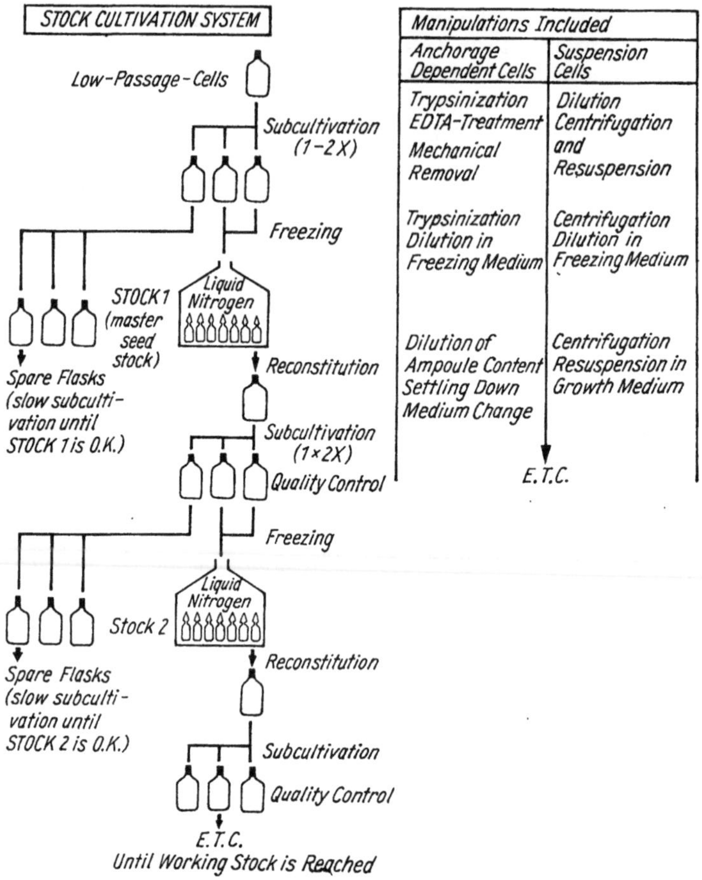

Fig. 3. Maintenance of cells using a stock cultivation system which combines security, uniform quality and an almost unlimited supplementation of cells with a minimum of space requirement.

separate place in order to prevent total loss of the cell stock by force majeur. Besides the nitrogen storage scheme, a simple room temperature storage is also recommended. Here a minimum of one subculture bottle from each cell strain should be stored in a dark place at room temperature (25—33 °C) as spare flask. This will remain useful for several weeks.
The maintenance of cultured cells requires many manipulations, more so for anchorage dependent cells than for cells growing in free suspension (Fig. 3).
A simplified growth scheme for the mass propagation of both types of growth using a baby hamster kidney cell line (BHK 21) as an example, is shown in Fig. 4. BHK 21 cells

which are initially anchorage dependent cell types can be grown in suspension (as can some other cell lines) by slight agitation of freshly passaged cells in siliconized culture vessels with special suspension media. Both types of growth can be obtained after this treatment.

It is very simple to obtain high cell densities from cells growing freely in suspension and need not be especially labour intensive. As pointed out in Fig. 4 quantities of the order of 10^{13} cells can be produced quite early in a weekly working rhythm. This goal requires a series of bioreactors each of successively larger volumes where the harvest of the respectively smaller vessel serves as the inoculum for the next. The tending of these bioreactors is quite simple, as the entire procedure is carried out in a closed system. Provided adequate equipment is available, a single technician could cope with this task.

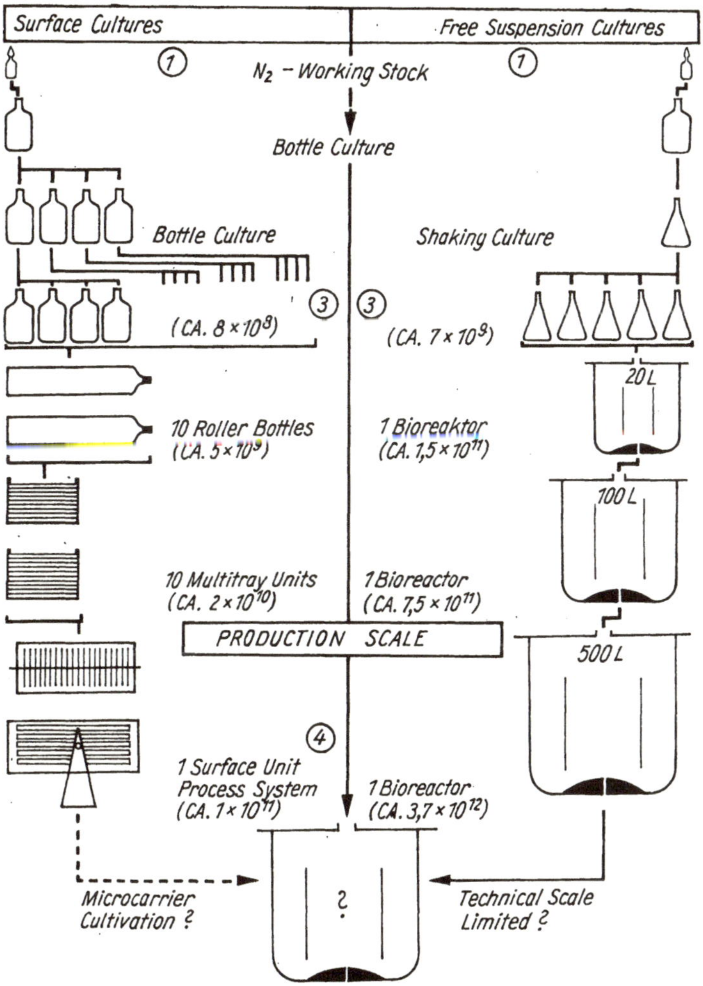

Fig. 4. Scheme of growth for the mass propagation of cells (BHK 21 as an example). (1) Maximum cell density at confluence is 3×10^5 cells/cm² according to own experience. (2) Maximum cell density is between $5-10 \times 10^6$ cells/ml according to [50, 117]. (3) Numbers in parenthesis indicate the total amount of cells produced in this stage, the same labour assumed for both surface and suspension culture resp. (4) The largest known surface unit process system contain a surface area of appr. 30 m² [30].

The largest commercial system currently available for the cultivation of anchorage dependent cells (multi-surface propagators such as the Gyrogen, Chemap AG, Switzerland, offer approximately 30 m^2 of surface for cell growth. In the case of BHK 21 (see Fig. 4) a total cell number of 10^{11} can be obtained, i.e. two orders of magnitude less than that obtained from an easily manipulated bioreactor of approx. 2000 litres working volume. Therefore, with respect to cell propagating capacity, about 100 large scale surface propagators are equivalent to one 2000 litres bioreactor for free suspension culture.
Of greater interest is the difference in the amount of work involved. To inoculate one of these large surface reactors alone, the cell harvest from 40 roller bottles, or that from 10 multi tray units is required [30]. It is evident therefore that, with the greater amount of work involved in trypsinsization and inoculation, and the consequently greater risk of contamination, the mass cultivation of anchorage dependent cells in surface propagation will, therefore, always be limited. Very early in the development of tissue culture, attempts were made to make the advantages of free suspension techniques accessible to the cultivation of anchorage dependent cells. Microcarrier culture (the growth or maintenance of anchorage dependent cells on small beads suspended in a bioreactor) provides a convenient means with which the problems of mass cultivation can be overcome.
The technique of microcarrier culture, first proposed and applied by Van Wezel [128], encounters many problems, which are also found with free suspension culture and will be considered in a later section of this contribution. The methods of surface culture have been the subject of a recent review [111].

4.2.2. Routine quality controls

The management of cells and the preparation of inocula for large scale propagation are potential sources of infection for both the laboratory personnel and the culture itself. The potential biohazard for the laboratory workers centres on the fact that cell cultures may carry unrecognized pathogenic viruses or are tumour derived. In this respect, primary cultures are probably more dangerous as potential sources of laboratory-acquired infections than are well-defined continuous cell lines which are assumed to have a zero or low level of human pathogen contaminants.
Some ingredients of cultivation media, (mainly serum) together with other additives and reagents used in handling the cells and the manipulation steps represent constant sources of potential contamination which will inevitably lead to failures in cell cultivations. Some types of contamination, such as fungi and bacteria, are easily detected since they usually grow rapidly and produce turbidity. Other types of contamination, such as mycoplasma or virus, very often remain undetected until a late stage of cultivation. Each laboratory therefore has to decide which controls are feasible for its own use. In our experience, the following standards of control are to be recommended for cell maintenance and the preparation of cell inocula for mass production in order to maintain a high level of reproducibility and the minimum contamination over a long period:

1) Antibiotic-free growth media should be used whenever possible, particulary for the preparation of inocula for large scale cultivation. In this way, contamination with bacteria of fungi is recognized at an early stage and cultures can be discarded in time. A special control for anaerobic microbes should also be made.
2) All the material necessary for cell cultivation should be brought in large quantities and subjected to the control procedures systematically presented in Table 4.
3) Growth media should be as complete as possible when they are sterilized and stored. In our laboratories, all growth media are prepared ready for use with serum and glutamine incorporated.

They are kept frozen until all the controls shown in Tab. 4 are completed. Media without protein are stored at 4 °C and unstable additives are added immediately before use. In our experience, adequate sterility control of growth media is achieved when the room-temperature-storage culture system outlined in section 4.2.1 is performed without antibiotics in parallel with growth potency and mycoplasma test.

Table 4. Routine quality controls performed in our laboratories

ROUTINE QUALITY CONTROLS

Media, Sera	Trypsin, Antibiotics, Additives	Flasks, working Material working Area
osmolarity	osmolarity	sterility
sterility	sterility	quality of surface treatment (plastic culture vessels)
mycoplasma contamination	mycoplasma contamination	toxicity (plastic and rubber parts)
virus contamination	virus contamination	germ absence (working area)
pyrogen content	pyrogen content	germ concentration (laboratory)
growth supporting potency	toxicity	residues of cleaning agents
purity (water)	efficacy	

↓ ↓ ↓

CELL CULTURE

PROTOCOL (genealogical register)
- lot numbers of media and material
- population doubling level
- passage level
- cell count
- viability
- mother-flask-number
- results of controls

CONTROLS for
- morphology
- sterility
- mycoplasma contamination
- virus contamination
- contamination with other cells
- karyology
- function desired

↓

PRODUCT

Specific Product Control

4) Mycoplasma in cell cultures [19, 106] may be identified according to the following methods: (a) the agrar/broth 4-step culture method according to BARILE [4], (b) the fluorescent DNA-staining method described by CHEN [16], and (c) the radio labelling-sucrose-banding method modified by TODARO [119]. b) and c) are used in parallel for suspension cells.
Growth media are routinely checked for mycoplasma by passaging control cells (1 diploid strain, 1 permanent line) four-times in the medium to be tested, and thereafter checking the cultures for mycoplasma contamination as described above. This procedure has been found to be more useful than other methods because of its higher sensitivity in detecting cell culture dependent species of mycoplasma. This test may usually be combined with an estimation of the growth promoting potency of the growth media as described below.

5) Virus contamination is routinely tested for during the incubation of control cultures by prolonged observation of cytopathic effects. Supernatants of cultures under test are also used to inoculate a variety of other cell cultures known to be susceptible to many viruses (e.g. Vero, HEp2, MRC5). These are then observed for at least two weeks, after which the susceptibility of these cultures to contamination is ascertained by challenging with a control virus.

6) Pyrogen content is detected by the LIMULUS amebocyte test [133]. This test indicates a former bacterial contamination of water, media or serum, which has been removed by sterilization. Precontaminated media may still be used with caution, provided that growth can be adequately supported. They should never be used, however, for the maintenance of stock cultures.

7) Growth supporting potential of sera and media is estimated by passaging a particular cell substrate for a minimum of 4 subcultivations. The composition of the medium should support the growth of more than 1 cell line of the most sensitive cultures (MRC5 and MDCK or Vero are used routinely in our laboratory). Reference test cultures with medium or serum of well known growth characteristics are subject to the same procedure of subcultivation. Growth curves are then constructed from the cell growth data under both media and these should be identical for each subcultivation. The relative growth potential is indicated by comparing the growth curves obtained with different media or sera. After the first passage clonal growth potency is estimated.
8) Toxicity test for antibiotics and other additives are done by seeding cells from a pooled cell suspension onto cluster plates and adding dilutions of the substance to be tested followed by incubation for 48–72 hours. Total cell protein is then estimated by staining with neutral red, extraction with alcohol and photometric measurement. Any growth inhibition can then be identified by plotting the results obtained and comparing these to reference substance curves.
9) All other routine quality control procedures shown in Tab. 4 are perforemd according to standard procedures.
10) Prolonged subcultivation of cell lines may lead to changes in some cell characteristics since neither the growth medium nor the growth condition could ever completely simulate the cell environment *in vivo*. For example, changes in growth patterns, loss of some specialized functions and loss of characteristic antigens may occur. Although of greater importance to licensed diploid cell strains, which are used in the manufacture of vaccines, the karyological status should also be periodically assessed for line cells. This is particularly useful for the diagnosis of aberrations caused by selective pressure, overgrowth by a subpopulation; persistent virus infections and cross contaminations with other cells. A great number of procedures (including automated flow cytometry [17]) are currently available for the preparation and analysis of many cell karyotypes. This type of work is most suitably carried out by experienced personnel.
11) Routine control is supplemented with controls directed towards specific cell functions, which are central to, and thus representative of, a particular product orientation.

The large quantity of work and costs incurred in all of these control procedures will be necessary as long as complex, serum-containing non-autoclavable media have to be used in cell cultivation processes.

4.3. Growth media

The basic goal in the design of culture media is to stimulate growth to high densities whilst maintaining the desired specialized functions of cultured cells during conditions of prolonged subcultivation; i.e. the growth medium should mimic the tightly regulated *in vivo* environment. Quite contradictory to these requirements, growth media for large scale cultivation must be cheap and autoclavable.

As shown in table 5 a variety of media compositions requiring no addition of serum to support growth have been published for the cultivation of many cell types (compare 5 and 52 for review). However, for a variety of reasons, the use of such serum-free culture media is quite uncommon. So far, more than fourty different formulae have been published on the basic media composition optimal for the growth of the wide range of animal cells currently being studied. Serum supplemantation is central to the design of well-balanced media as it contains many well-known components in addition to the numerous unidentified ones in minute quantities which are often essential for good culture development. Since the composition of serum is variable and dependent on both its source and preparation, its supplementation into these media is generally excessive (usually a 10%(v/v) is added) in order to ensure adequate availability of these components to the cells.

The risks that, during prolonged subcultivation in serum-free media, special cell features such as antigen production, membrane receptors, virus susceptibility, etc. may be

Table 5. Cell lines cultivated on serumfree media

Cell Line	Medium	Remarks (Supplements)	Reference
Rat follicular	DME/F12 Mixture	Hormones and Growth Factors	13
Swiss 3T3			
Balb/c 3T3			
Mouse Melanoma			
Mouse Testicular			
Rat Glial			
Mouse Carcinoma			
He La	F12	Hormones and Growth Factors	40
MDCK	DME/F12	Hormones	116
Lymphoblasts	DME	Albumin, Transferrin, Soy bean lipid	42
Mouse Myeloid Leukemia	F12	Trace Elements, Albumin, Hormones	38
Epithelial Cells	Diff. Components		125
L929	DME/F12	Steroids, Hormones	48
Hepatoma	DME	Tripeptide	96
Mammary Tumor	EMEM	Diff. additives	7
Py BHK, SV 3T3	MB 752/1	Iron	14
BSC-1, 3T3, CHO	MEM-α	Peptone, Insulin	54
3T3, Hela, KB BHK 21	GFAD	—	120
Lymphoblasts	FCRC-1	—	85
Mouse Mammary Tumor	FCRC	Hormones	86
REVIEW			5

lost, is usually regarded as being very high [31, 53]. Therefore, in spite of its manifold disadvantages, the supplementation of media with serum remains the simplest alternative to the use of defined compositions. In large scale mass culture, the use of serum enriched media is still commonplace although the problems arising from this are much more serious. The main shortcomings are these:

- Serum is the primary cost factor in large scale cell production, regardless of whether it is supplied by a commercial source or from within the company itself. On average, the supplementation of medium with 10% serum accounts for 80% of the material costs;
- Serum represents the major source of contamination from viruses, mycoplasmas, bacteria, etc. [4, 68];
- It is difficult to obtain very large quantities of serum of constant quality and composition. This means that expensive and time-consuming quality control measures are necessary. Therefore, it would seem economically and scientifically wise for companies to obtain a large supply from one source at one time which would be sufficient to satisfy the cultivation requirements for at least several months. Provision of such large amounts of serum, particularly if foetal sera are necessary, requires the use of an expensive service organization;
- Serum, per se, represents a pyrogenic contamination which highly complicates the recovery and purification of the final product;
- Certain inhibitors in the serum, such as γ-globulins can act unfavourably on the production of certain biologicals (e.g. virus replication or urokinase production).

In the light of these aspects, one key to a more economic animal cell technology undoubtedly lies in the reduction or, even better, in the complete replacement of serum in media formulation.

Cheap autoclavable culture media which are sufficient to support the growth of a wide range of cells to high cell densities at reasonable growth rates, and which keep the cells unchanged in their phenotypic expression over an extended period of cultivation, would be optimal in this respect. The complete chemical definition of those media would be desirable, although not essential. Naturally, those substances which are intended to replace serum, either partially or entirely, must be cheap, otherwise the direct serum costs cannot be substantially lowered. The ability to autoclave such media would further reduce the indirect cost factor of quality control.

All teams attempting the partial or complete replacement of serum in growth media are confronted with the same problems: (a) Serum can promote cellular growth in many different ways. Ham and McKeehan [32] have been able to list at least 34 possible ways in which serum may promote cell replication in conventional culture systems; many of which can now be regarded as being firmly established. However, many other functions of serum still remain to be identified. The necessity of replacing these unidentified functions by modification of the medium or other environmental parameters raises problems that can only be overcome by trial and error. (b) Different cell types have different nutritional requirements. Non-adapted diploid cells require different relative and absolute concentrations of single substances for optimum growth compared to typical permanent line cells, which have undergone adaptation *in vitro* during extended subcultivation. Line cells tend to have less specific growth requirements compared to diploid cells and thus a reduction in, or a total replacement or the serum requirement is easier to achieve line cells. (c) The nutritional requirements of a given cell type varies with the cultivation system employed. Major differences between culture systems are dependent on cellular population densities, on whether the cells are grown attached to a support surface or are in free suspension and also on whether the cells are kept in a static or in an agitated system [53]. A medium composition which has been found to be optimal for a given cell type in a given culture system is not necessarily optimal when the culture system as a whole, or the scale of operation alone, is changed. Therefore medium compositions often have to be redesigned when such changes are made.

Considering the obvious requirement for more knowledge on the nutritional demands of cultured cells and the optimization of the media, there are comparatively few small research groups working on these problems. These groups can be divided into two categories:

a) Those who are carrying out purely fundamental research in that they systematically explore the basic nutritional requirements of diploid cultured cells for growth and function. They partly work with strictly defined cultivation models under the motto "alter the medium, not the cell" [63, 70, 71].

   It is the goal of these teams to define the growth requirements on a broad basis. Although they have not yet fully reached their objective, the work has at least come to a point where the use of better-defined and well balanced formulae, especially with regard to trace elements, hormones and specific growth factors has allowed the serum requirement for some diploid cells to be reduced to a minimum [69].

   Other teams involved in basic research are primarily working on permanent cell line models following analogous objectives. Their work has led to the establishment of serum-free media for many cell lines [5, 42, 48, 130]. However, the expectations of the industrial expert searching for inexpensive media for industrial scale production are still not fully met, since the addition of a variety of growth factors may often lead to media which are even more expensive than those supplemented with serum.

b) The strategy of those doing typical applied research is principally aimed at replacing serum with cheap, but often chemically undefined substances, or at obtaining technically practical media by other means.

   Simple commercially available media supplemented with peptones (cell enzymatic digests) or complex mixtures of peptones, together with serum albumin fractions [81], bovine albumin [9, 48, 87], unsaturated fatty acids [35, 134] or with peptone and hormones [34, 44, 54, 76, 84] have been shown to support continuous growth of several line cells such as human lymphoblastoid lines, Chinese hamster ovary, baby hamster kidney, mouse L-cells etc. in static culture. In agitated suspension culture, methyl cellulose or "plasma expanders" were found to be necessary to protect cells against mechanical damage. An elevated requirement for iron, selenium or other trace elements has also been observed for some cells cultivated in agitated or air-sparged systems.

Another successful approach in the formulation of cheap culture media for industrial scale processes, is that of using cheap by products of human plasma fractinations [55, 70]. Many protein components which have been identified to promote growth in serum-free media (insulin, somatomedin, transferrin, ceruloplasmin) have already been recovered from by-product fractions. Treatment of "cheap sub-standard serum" (bovine serum, human serum etc.) with polyethylene-glycol in order to remove specific and non specific antibody activity as well as contaminating viruses and mycoplasmas [6, 41] has practical advantages for both sterile filtration of media and final product purification, in the large scale production of viral vaccines and other biologicals.

It is certainly not yet possible to formulate culture media which are fully ideal for large scale use from the methods of optimization discussed in this article. Nevertheless, it can be seen that it is possible nowadays to produce inexpensive media, which meet almost all of the criteria of an industrial medium. The future replacement of those serum constituents, which have been identified by basic research as being critical for cell cultivation with cheap substitution compounds, are a major prerequisite for the industrial advancement of animal and human cell technologies.

5. Large Scale Cultivation

What does "large" or "industrial" scale mean for animal cells and what are the upper and lower limits? It is difficult to answer this question precisely. Factors such as the nature and demand of the product, the economic risks involved and, more especially, the technological problems, must all be considered together if an optimal scale of cultivation is to be achieved. The mass culture of cells alone represents the major bottleneck in the process design. Some examples of industrial scale processes soon to be, or currently in operation, may be used to show the scale of production of biologicals which is being carried out with permanent cell lines in suspension culture.

Several pharmaceutical firms are going to industrialize the production of interferon, perhaps "the drug" of the eighties and nineties. To secure a supply for exhaustive clinical trials, which will take several years, semi-commercial production plants are needed. Using the data published by Klein [61] as a reference, approximately 20.000 litres of cell culture per year would be needed to treat 100 cancer patients over the same period of time. In a batch process (i.e. approximately 35 weeks of operation p.a.) this would require a "growth vessel" with a working volume of 600 litres. Although this number of 100 patients is negligible compared to the target number for this drug, the unit mentioned above is considered as "industrial scale".

There are now commercial units for the production of FMD-vaccine operating with culture vessels of more than 1000 litres working volume. Large vaccine factories in South America now coming into operation have design capacities of up to 600 million monovalent doses per year [82]. A weekly harvest of approximately 20.000 litres of cell culture, BHK 21 grown in suspension culture, will be required to produce the FMD virus for this purpose.

Integrated crop pest control programmes using viral insecticides would require production units of several orders of magnitude greater than those mentioned above. According

to estimations published by POLLARD and KHOSROVI [99], 7.5 t of dried virus infected cell mass, annually produced by a series of two storage continuous bioreactors of 200 m^3 total working volume, are required for the treatment of an area about 10^6 acres in size. Potential biohazards associated with the growing and processing of large quantities of cells or viruses and the handling of large quantities of expensive material with high risk of contamination, etc. have to be met by special architectural and infrastructural measures which exceed the scope of those so far undertaken in biotechnology [2].
However, in this communication only the problems of biochemical engineering are under consideration and these become most significant for large scale processes of propagation.

5.1. Biochemical engineering considerations

Several unit processes are involved in large scale cultivation (compare Tab. 6). The larger the scale of operation, the more important is the ability to carry out the steps of the operation in a functionally and economically satisfying way. In addition to actual mass cultivation, the processing of water, the sterile filtration of media, and the cell harvest (while maintaining full viability) all deserve special attention.

Table 6. Unit operations involved in large scale propagation of cells (minor components omitted)

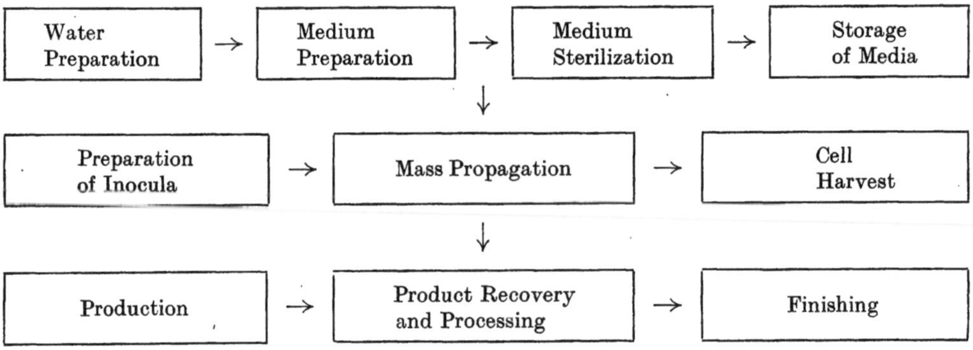

5.1.1. Water preperation

It seems trivial to devote space to a simple question such as water processing. However, from many personal communications the authors are of the opinion that inadequate water quality represents a frequent source of trouble, and that the consequences arising are particularly unpleasant. Water is often only recognized as the causative factor in industrial problems when the entire process, from the strain propagation to the production, has come to a halt, or when inexplicable changes in cell growth are observed. Substandard water quality is, in most cases, masked by the detoxifying properties of the serum. Thus, water quality is of critical importance especially for serum-free or low-serum media. It has become a common practice in many institutions to use 2—3 times distilled water for the preparation of media. According to our experience this is a luxury which is not economically justifiable in industrial-scale applications.
The quality of the Viennese drinking water is often adequate for cell culture, provided a non-defined complex medium is acceptable. The same water can, however, by a badly attended demineralization unit and inadequate storage, contain pyrogens to an extent which makes it unsuitable for cell culture. In the case of having drinking water quality at one's disposal, it is usually sufficient to a) clear the water of organic matter by ultrafiltration or by reverse osmosis b) demineralize it by ion exchange methods, or more suitably by distillation and c) store it at 80 °C to prevent bacterial growth.

It is very important to correlate water processing capacity with the demand in order to avoid long storage and to permit cleaning of ion exchange cartridges and reverse osmosis equipment once a week. The steam necessary for direct sterilization of vessels and other equipment should always be produced from the same water free of any additives in heat exchangers made from stainless steel.

5.2. Mass propagation of cells

Mass propagation in the growth vessel is central to every production process concerning animal cells. This unit process determines the mass and quality of cells and thus the efficiency of the entire technology. Bioreactors similar to those originally developed for the submerged cultivation of microorganisms are currently in use for the cultivation of free suspension culture as well as for the cultivation of cells on suspended micro-carriers. Modifications of these microbial reactors are necessary to meet the fundamentally different requirements of animal cells. With this in mind, it should now be possible to cultivate cells in unit processes up to a volume of several thousand litres. With bio-reactors of several hundred to over 1000 litre working volume there are usually no exact data available on important biological parameters such as growth rates, cell densities, etc., which would make a biochemical engineering analysis on a quantitative basis possible. Often it is not clear whether such large reactors really function as growth vessels or are reactors in which no substantial propagation takes place. Examples of this latter situation would be where the cells are kept in a stationary growth phase for use merely as the substrate for virus replication or for the production of a specific metabolite.
To some extent it is also true that many of the parameters, although currently accessible, are not being measured. Owing to this situation that biological scale-up concepts have not been discussed or tried out on suspension cultures (or microcarrier cultures) of animal cells to the extent indicated by the importance of the problem. Therefore, we would like to try to identify on an abstract basis, as far as possible, the parameters relevant to the mass culture of cells. Following on from this, we will then attempt to discuss possible ways of scaling up reactors for the suspension culture of animal cells.
The importance of the animal cell is an inherent feature of all our considerations. We require it to express its full biological potential as influenced by both chemical and physical manipulations of the environment. The major drawback, however, is that we know little in qualitative terms, and even less in quantitative terms, about the environmental requirements of these cells. It therefore becomes very difficult to set out methods on how to carry out environmental manipulations, as well as stipulations on which measurable parameters are the most important and in need of control. Fig. 5 attempts to inter-relate qualitatively the range of factors affecting a culture. Culture development *in vitro* must be seen as an indivisible event in which the final result, i.e. the "culture", is determined by an interplay between the cells on one side and the combined chemical and physical input on the other. The culture's activity feeds back upon its development by rheological and/or chemical feedback loops (e.g. the intense foaming which arises in certain phases of culture development, the minimum inoculation cell density which is necessary for growth, the situation in which growth frequently stops at certain densities etc.).
The physiological state of the culture, and the question of whether a culture procedes according to, or deviates from, subjective expectations, can be judged by assessing a number of variables by means of the available set of analytical tools. These then allow us to draw conclusions about the reaction at morphological and chemical levels.
Only recently have some of these variables been recognized as being significant in the development of a culture and these become useful as parameters [117, 118] in preventing uncontrolled, spontaneous developments of the cultutre. Some of the parameters —

shown in Fig. 5 supply us with information on the extent to which physical input or chemical factors positively influence the culture. Theoretically, these could be used for the interactive control of culture development, either via programmed feeding of nutrient media (or of single components) or via operational measures (perfusion culture, temperature programme, etc). The problem is that there are still no rapid and precise methods of analysis suitable for use in either *in situ* or in bypass loops which deliver the appropriate sensor signals. In most cases there is a lack of analyses which could qualitatively correlate the available sensor signals with the metabolic events occurring in the cells. The proposed application of computer control to animal cell suspension cultures [89, 110] are, for these reasons, of limited value. In reality, only a few parameters are currently controlled via closed loop systems (compare Fig. 5).

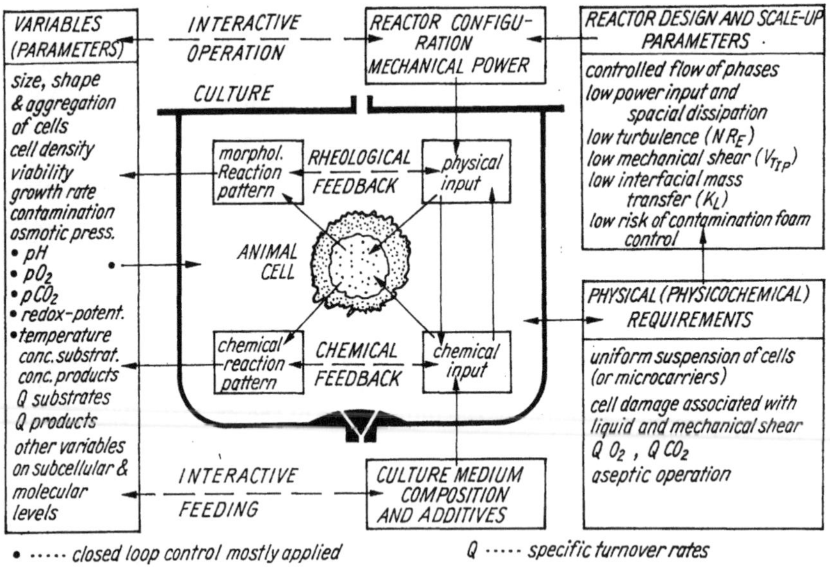

Fig. 5. Illustration of in vitro animal cell culture interrelationships: Reactor design parameters and measurable/measured and controllable/controled variables — an abstraction.

The need for more intensive development and application of sensing techniques has already been recognized, particularly in connection with the direct determination of various metabolic functions related to cell differentiation. If adequate sensors could be implemented, interactive control would result in environmental conditions which would more closely approximate to those from which the cells originated (compare Chapter 5.3).
Reactor and scale-up concepts are most easily applied when based on well-defined data. As far as animal cell culture is concerned, there is only a limited amount of data available. Very few physical (physicochemical) requirements, which are needed for the design and scale-up of bioreactors, are qualitatively or quantitatively identified as being important (compare Fig. 5). Uniform suspension of cells (or cells adhered to microcarriers) with a minimum of cell damage, is one of the major requirements. Liquid (turbulent) shear is most probably as detrimental to the cells as is mechanical shear (shear results from a variety of interactive forces within the system). A few quantitative data exist which may be used to characterize the interfacial mass transfer capacity which a bioreactor must possess for the successful cultivation of cells. In particular, such transfer must be able to satisfy the specific oxygen demand of animal cells in culture. Table 7 shows a comparison

of the specific oxygen uptake rates (Q_{O_2}) of 4 different cell types under various conditions of cultivation. It is evident from these values that the specific oxygen uptake rates correlate well with the specific growth rates irrespective of whether the cells are grown in free suspension or exist on microcarriers. The critical oxygen tension (pO_2) for all of these cell types is around $<40\%$ of air saturation. This means that if the pO_2 drops below this value the specific oxygen uptake rates and the specific growth rates will be reduced.

Considering the above information, therefore, the average oxygen transfer rates for a typical animal cell culture reactor should be in the range 0.1 to 1.0 mmol $\cdot$ $l^{-1}h^{-1}$. An oxygen supply of 0.1 mmol $\cdot$ $l^{-1}h^{-1}$ adequate for a cell density of 10^6 cells per millilitre. Typical "microbial reactors" have designed oxygen transfer capacities between 50 and 300 mmol $\cdot$ $l^{-1}h^{-1}$, which are several hundred times higher than would be suitable for animal cell culture. In the light of these different requirements, the question which must be seriously asked is whether microbial reactors would really solve the problems inherent in animal suspension culture.

5.2.1. Reactor design parameters

The various reactor configurations which have been used for free suspension (or microcarrier) culture of animal cells may be divided into two basic categories, the mechanically agitated reactor and the pneumatically stirred type [50]. Most of the mechanically agitated reactors are of the Rushton-type configurations (i.e. a stirred vessel of fixed geometric proportions).

A few reports also describe the use of the Chemap-Vibrofermenter for suspension culture [78, 28, 50, 129]. It would not be within the scope of this contribution to summarize all the available information concerning reactor configuration. However, we would like to try to point out those principal factors which are essential to our design considerations. These are summarized in Fig. 5:

a) Controlled flow of all phases in (and around) the reactor is necessary in order to maintain the cells (or microcarriers) uniformly suspended and to prevent special gradients of reactant concentration from being established. Streamline flow and low mean liquid velocities (a few centimetres per second) are the preferred flow characteristics. Interfacial mass transfer rates, and especially oxygen transfer rates, have to be properly adapted to the low requirements of animal cells.
   Total power input, local power dissipation densities and turbulent and mechanical shear rates, all have to be kept at the lowest possible level, otherwise cell damage may occur. This is particularly true for low serum and serum-free culture media, since the macromolecular compounds of serum afford some protection to the cells against physical damage. Although other macromolecules can partly replace this function of serum [9, 53, 84] such negative effects on the cell growth should *a priori* be eliminated whenever possible.

b) Most of the research and development information on cultured cells has been gained from work carried out on millilitre- or litre-scale. Scale-up of a reactor to the cubic metre-size will result in altered physical input variables, since it is not possible to keep the physical functions of a reactor constant as regards geometric kinematic and dynamic properties. Changes in the physical system may lead to unexpected responses by cultured cells, either from the point of view of overall growth or from changes in their subcellular properties (e.g. loss of certain surface antigens as shown by Zwerner et al. [136]). Anyone attempting to scale up a process based on the growth of cultured cells must therefore be prepared for unexpected events. When such physical changes can be controlled during the scale up of reactor functions, the risk of unexpected changes occurring in the expression of desired physiology will be reduced. Simple and transparent reactor systems, with respect to their functions, are therefore to be preferred.

c) Animal cell culture media are rich in nutrients and, because of the consequent low selectivity, are unable to suppress the growth of contaminant bacteria etc. Furthermore, due to high media costs, contamination is a grave economic burden. For these reasons one should choose reactor types which are particularly safe in this respect. Again, this leads to the preference for simple technical construction for cultivation purposes.

5.2.1.1 Reactors for suspension (and microcarrier) culture

Most of the above parameters are met by the idealized cycle flow or liquid bulk flow pattern sketched in Fig. 6A. This design would also be most desirable with respect to minimum shear and to the attainment of uniform suspensions. Bulk flow regimes occurring in biological reactors have been intensively analyzed on a theoretical [51, 102, 122] as well as on an experimental basis. All of the reactors designed on the principle of recycle reactor have one feature in common; scale up is less complicated to perform than with stirred vessels.

The RUSHTON-type reactor configuration (stirred vessel Fig. 6B) is obviously the one which deviates most from the desired concept. Nevertheless, the stirred vessel is the most frequently used bioreactor for both microbial processes and suspension culture of animal cells. When it is applied to the cultivation of animal cells, this configuration has to be redesigned to meet their specific requirements.

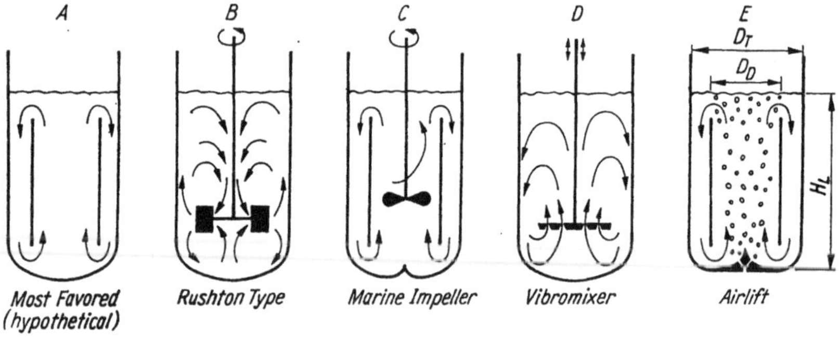

Fig. 6. Liquid flow patterns. (A) Desired regime of flow for the cultivation of animal cells in free suspension (or on microcarriers). (B) Bulk flow pattern in the Rushton type reactor. (C) Axial flow pattern generated by marine impellers. (D) Vibrofermenter. (E) Airlift-type reactor.

Design problems associated with this reactor configuration arise mainly from the facts that its liquid bulk flow pattern is not streamline and that all of the numerous physical functions which the reactor is required to possess, depend on the geometric configuration and agitation speed of the stirrer (i.e. the stirrer is multifunctional and therefore complicated in scale up). Typical scale up behaviour of the stirrer are such that, with increasing size but constant geometric similarity, bulk flow behaviour becomes diminished, whereas turbulent and mechanical shear fields concurrently tend to increase [90]. This situation is totally antagonistic to the aims of animal cell culture since it has been shown that liquid and mechanical shear forces often quantified by key dimensionless numbers, are responsible for the disintegration of emulsified or suspended particles. Such forms also have damaging effects to cells above certain limits which are as yet undefined quantitatively [90]. Therefore, maintenance of geometric reactor similarity, often practiced for the scale up of conventional stirred tank reactors, is not reliable, since the functions served by fast-moving small mechanical agitators dominate over the effective momentum transfer which is necessary to produce liquid bulk flow. This phenomenon of mechanical agitation can partly be overcome by an increase in the D_i/D_t ratio and by a decrease in the agitation speed in scale up systems. The scale up of agitators, based on the criteria

of momentum transfer, is discussed in more detail by BLAKEBROUGH and SAMBAMURTHY [11].
Reactor configurations which use flat blade turbines for agitation (compare Fig. 6B) are generally inadequate. Mechanical agitators producing axial liquid flow such as the marine-type propeller or vibromixers may achieve conditions which approximate to the physical requirements of animal cells during submerged cultivation as may be concluded from theoretical considerations. A certain amount of shear caused by the high local power dissipation density which occurs in the region near the impeller tip or around other mixing devices, still remains. Whether or not these shear fields are detrimental to fragile cell tissues has yet to be investigated using empirical trial and error approaches.
It appears that the theoretical design requirements for an animal cell reactor, mentioned in 5.2.1 are most easily met using the airlift-reactor concept. The operating principle of such a reactor is presented in Fig. 6E. A gas stream is introduced through orifices into a cylindrical vessel fitted with a draft tube for directing the liquid flow into a controlled recirculation stream. As can be seen from a detailed analysis of this reactor type carried out in our laboratory [KATINGER et al., 50], this design principle provides the reactor with most of the characteristics required by a suspension culture reactor for animal cell cultivation, i.e.:

a) Only minute amounts of gas need to be introduced in order to promote controlled flow of the bulk liquid at the minimum speed necessary to maintain cells in suspension (the mean liquid velocity for this is in the order of 3 to 10 cm · s^{-1}).
The formation of bubble swarms is prevented by this mode of operation and therefore, problems with foam production are avoided. Consequently, a 100% degassing effect is achieved on the liquid surface and for this reason, the power required to bring about liquid motion is mainly derived from the difference between the apparent densities of the contents in the gassed (ascending flow) and in the ungassed (descending flow) section of the reactor.

b) The scale up of this reactor is relatively easy to achieve since it reflects, in reality, the mathematical abstract principles of a recycle reactor (compare [102] for detail) very closely. The reactor should be scaled up using the concept of keeping both kinematic and dynamic aspects as close as possible to the original design. The parameters for the airlift reactor are: aeration speed introduced into the system, the draft tube diameter (D_d)/tank diameter (D_t) ratio and the ratio of (D_t)/liquid height (H_L). Figure 7 shows examples of the qualitative and quantitative interrelationships of these parameters.

c) Reactors properly designed on the above principles for animal cell culture purposes have an extremely low power input per unit volume (P/V). The maximum (P/V) value which has been measured by us is in the order of 10 to 15 watt · m^{-3}. Obviously the local power dissipation density in this reactor must be very low, since not only is the total power absorbed extremely small compared to other types, but it is also specially dissipated over many air bubbles. The overall action of the air bubbles in contributing to liquid bulk flow is thus comparable to many small agitators. The ascending velocity of gas bubbles relative to their liquid environment is a function of bubble diameter and liquid viscosity and can attain a maximum value of 30 cm per second. In comparison, the shear rates at the gas/liquid interphase of ascending gas bubbles are of similar orders of magnitude to the shear rates occuring at the stirrer tip of small spinner flasks (200 ml size) stirred at 200 rpm. We can conclude from these considerations therefore that the shear rates created in an animal cell culture reactor designed on the airlift principle should not be more detrimental to the cells than those found in spinner cultures.

d) The airlift reactor, when it is operated under conditions appropriate for animal cell cultivation, displays a volumetric oxygen transfer rate (OTR) of between 0.2 and 1 mmol $\cdot l^{-1}h^{-1}$ [50]. These oxygen transfer rates, which can be obtained on any desired scale of operation, precisely meet the oxygen demands of animal cells, as seen from Table 7.

The airlift type reactor (or, more precisely, the bubble column reactor) has been successfully used for the suspension culture of several cell lines including BHK 21, lymphoblastoid cell lines such as NAMALWA and REH and a mouse/human hybridoma cell line used in our laboratory. Cell viabilities of more than 98% and cell densities of up to 10^7 cells per millilitre were achieved with lymphoblastoid cells in continuous culture using low serum (2%) media. Direct sparging of air to control the oxygen supply has however been reported to cause cell damage in low serum containing media [58, 117]. This could be prevented by either increasing the serum concentration or by adding a nonionic surface active polymer. This type of sensitivity shown by cells seems to be a wide spread phenomenon in suspension cultures grown on low-serum media. Similar adverse effects

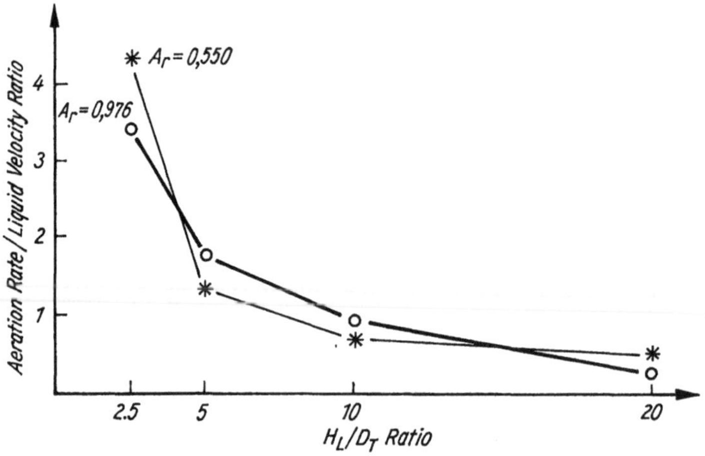

Fig. 7. Aeration rate/liquid velocity ratio against liquid height/tank diameter ratio in an airlift reactor of two different ratios of draft-tube area to downcomer area (Ar).

due to direct gas sparging were also evident in microcarrier culture systems, where sparging of air, according to our own observations, resulted in a reduced growth. In all these studies, the observed effects became evident several days after inoculation of the microcarrier culture, irrespective of the type of microcarrier used.

Thus, a variety of sublethal effects can be observed in different cell culture systems which are agitated and aerated by a variety of different methods. The complexity of this situation has so far prevented a detailed understanding of the major control mechanisms responsible. We must therefore reconsider the possibilities of optimally aerating and agitating these systems.

Several methods of supplying oxygen to aqueous culture media are possible: (a) direct sparging of gas, (b) head space aeration through the liquid surface, (c) oxygen saturation of media prior to its input into the reactor and (d) permeation through membranes and other more theoretically applicable methods.

Direct sparging of gas is the most often used method of oxygen supply. From the experimental information currently available on animal cell culture we can conclude that some cells (as in some systems such as those involving microcarriers), appear to be more sensitive to the presence of gas bubbles than others. The reasons for this differential sensitivity are not yet well understood. From a purely physical point of view, several mechanisms could be responsible for these observed negative effects on cell growth. It is conceivable that cells which are embedded in the air bubble liquid interface might be ruptured by bubble collisions as they leave the liquid surface or by coalescense and brake up of bubbles in the foamy phase above the liquid surface. Shear, due to velocity gradients in the fluid film adjacent to the gas/liquid interface are obviously not the dominating mechanism, as has been discussed earlier in this review. The method of gas sparging appears to be important. When air sparging is applied in a mechanically agitated reactor, the shear fields existing near the agitator produce bubble swarms of very broad size distribution. The rate of bubble coalescense in cell culture media is not correspondingly fast enough, so that air sparging in those systems results in severe foaming with cells

Table 7. Specific Oxygen uptake rates of different animal cells growing in different cultivation systems

Cell Type	Cultivation Technique Applied	Specific Oxygen uptake ($Q\ O_2$) (μmol $O_2/10^6$ cells × hour) × 10^{-3}	Specific Growth Rate μ (h^{-1})	pO_2 mm Hg	Reference
BHK 21c 13 Baby Hamster Kidney	continuous Culture	85	0.018	60	50
		110	0.033	60	
Human Lymphoblastoid Cells (Namalwa)	continuous Culture	54	0.012	50	50
		75	0.020	50	
Mouse L 929	Microcarrier	20	stationary Phase at Confluence	60–80	49
		65	exponential Growth Phase	ca. 60	
Hum. Foreskin FS-4	Microcarrier	50	0.010	ca. 60	124
		85	0.025	ca. 60	

trapped within the foamy phase. In an airlift reactor, large and non-coalescing air bubbles of uniform size distribution are created. As a result, foaming problems do not usually exist in this particular system.

In this context, several physical aspects associated with gas sparging and some experimental phenomena observed in cell culture systems, suggest that a large fraction of the cell damage might be due to the amount of energy which is immediately set free either by bubbles bursting as they leave the liquid surface or by their destruction in the foamy phase. Surface tension and forces resulting from bubble break up are higher when small bubbles are produced. Proper operation of an airlift reactor results in large bubbles with lower surface tension. The addition of surface active agents also results in a similar effect, in that the cohesive forces at the gas/liquid interphase are reduced, thereby resulting in the formation of bigger and less stable (more elastic) bubbles. As a consequence of this, the liquid films surrounding the gas phase in a bubble tend to become thinner (monomolecular) before eventually bursting. As the thickness of such liquid films decreases, the possibility of cells (or other particles) becoming static due to their attachement at the liquid/gas interface is markedly reduced.

Both the free suspension culture of cells in airlift type reactors and the addition of surface active agents to gas sparged and agitated systems, have been observed to be beneficial to cell growth.

Other physical phenomena, however, might be responsible for the rather modest growth results which have been obtained with cells grown on microcarriers in the airlift-type reactor or with those grown in other systems using direct gas sparging methods. All types of microcarriers available from commercial suppliers are ball-shaped and have diameters between 50 and 200 microns [8, 24, 88, 95]. For purposes of rapid sedimentation, these microcarrierbeads are more dense than the surrounding liquid. These microcarrier beads are not as strongly subjected to molecular forces as are the much smaller singly suspended cells. Therefore, they have a strong tendency to gyrate when brought into a moving liquid phase. From a purely physical point of view, all these factors (i.e. the spin drift of the microcarriers, their higher densities and their lower surface/volume ratio compared with a single cell) must lead to higher velocity gradients and thus to increased shear forces encountered by the cells when they are attached to the surface of microcarrier beads. Again, an enhanced growth effect in microcarrier systems can be achieved when the viscosity of the culture medium is increased by the addition of serum albumin or by polymer blood plasma expanders (e.g. polyvinylpyrrolidone). We conclude from the above considerations that the use of rodshaped or planar microcarriers instead of spherical ones could improve the microcarrier culture system with respect to purely physical interactions.

Head space aeration through the liquid surface is the most commonly applied method of oxygen supply for animal cell cultivations in laboratory-scale vessels. Fundamental data, however, show that this method of oxygen supply is not sufficient on a large scale, since the liquid surface area/volume ratio decreases with increasing vessel volume. As shown in equation (1), the total oxygen transfer rate into the liquid (OTR) is related to the rate of oxygen diffusion through the gas/liquid interphase (K_L) to the area (A) of the gas/liquid interphase available and to the difference between the oxygen concentration in the liquid bulk (C_L) and the hypothetical saturation concentration of oxygen (C_S), which is in equilibrium with the corresponding gas phase.

$$\mathrm{OTR} = K_L A (C_S - C_L) \qquad (1)$$

K_L is related to the fluid motion at the liquid surface. Under the conditions equivalent to suspension or microcarrier cultures (e.g. a spinner flask of 200 ml liquid volume, agitation speed 200 rpm.), K_L-values of between 0.2 and 0.3 cm · min^{-1} are found, depending on the medium composition. From these K_L-values, and from the data shown in Table 7, one can calculate that head space aeration will become limiting when the size of cylindrical-shaped growth vessels exceed 30—50 litres (using a D_t/H_L' ratio of 1). In hemispherical growth vessels with vibrostirrers such as those used by MOORE et al. [78], 200 litres-scale will be the limit for head space aeration. Pure oxygen at elevated pressures in the head space of the reactor would improve head space aeration. Nevertheless, on larger scales, head space aeration is not sufficiently effective to meet the oxygen demand of animal cells growing in free suspension or on microcarriers.

Permeation of oxygen through membranes in addition to head space aeration, is another possibility for improving oxygen supply to systems in which the direct sparging of gas is not possible. However, form data obtained in our laboratory we can conclude that this method of oxygen supply is at present also problematic. Since the membranes used in the reactor for supplying oxygen have to be mechanically stable, a minimum of six kilometers of silicon tubing (one millimeter inner diameter and one millimeter wall thickness) would be necessary to aerate a reactor of 1 m^3 culture volume, when air is used as the oxygen supply. However if more gas-permeable and mechanically stable membrane material were available, gas permeation through membranes could become a reasonable alternative.

Saturation of media with oxygen prior to its input into the reactor is not reliable for free suspension culture techniques. Taking into consideration that air saturated media contains approximately 6 mg of oxygen per litre at 37°C and that the oxygen tension in the culture should be maintained close to 40% of saturation, it is necessary to renew the reactor volume 1 to 10 times per hour with freshly oxygen-saturated media. This would be possible for microcarrier systems in which the separation of carriers from the media could be more easily effected to enable reoxygenation and recycle of the cell-free culture fluid.

5.3. Choice of proper cultivation method

It has been found that the principles of continuous culture derived from the study of microorganisms are to a large extent also applicable to the cultivation of animal cells. For example, when mouse leukemia L 1210 cells were cultivated under glucose limitation in a chemostat, steady states were established which were characterized by constant cell numbers and volumes and by stable levels of DNA and RNA, etc. [121]. In our own laboratory we were also able to establish steady states under glucose limitation with the lymphoblastoid cell line Namalwa using low serum media (compare Fig. 8). Other permanent cell lines, such as BHK 21, Reh 6, hybridomas etc. can also be grown under steady state conditions (Katinger et al.; unpublished results), although with these, the nature of the limiting substrate during the continuous cultivation could not always be ascertained. Nevertheless, our own results and those of other workers [72] lead us to the

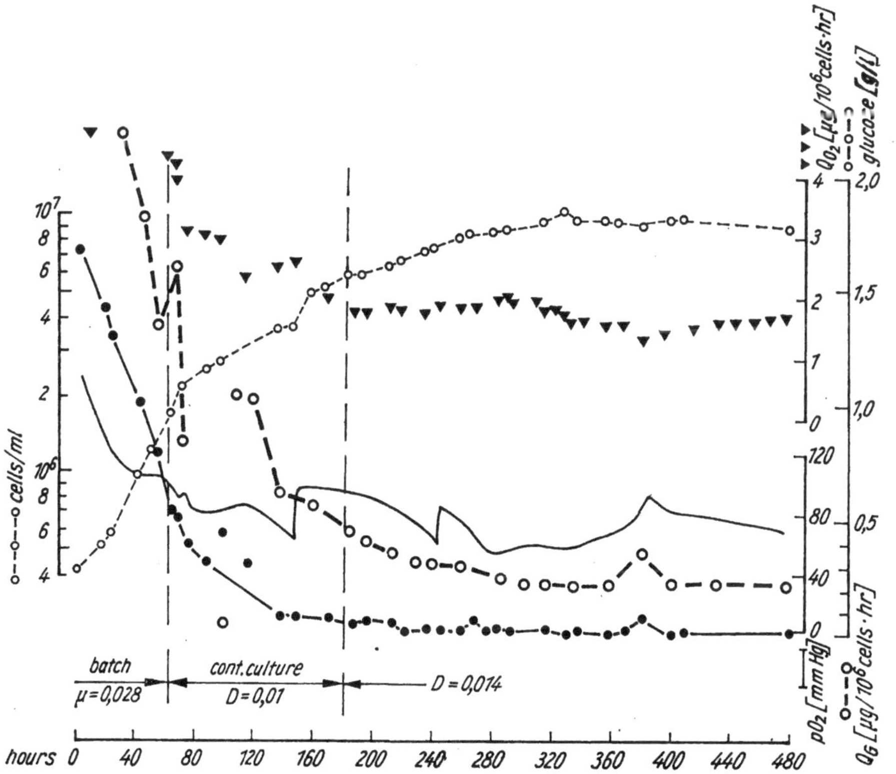

Fig. 8. Comparison of batch and continuous culture of Namalwa in RPMI 1640 + 2% human serum.

conclusion that simple alterations of the method of cultivation from a batch (closed) system to a continuous (open) system will always result in increased yields of cell mass. This phenomenon of enhanced conversion efficiencies of substrates into cell mass in open systems compared to closed systems is probably of general validity to all permanent cell lines.

According to our own results, cell yields in continuous culture of the Namalwa line were up to five times higher than those found in batch cultures using the same culture medium compositon. In this particular case, glucose was the growth-limiting substrate during continuous culture and a clear relationship between the specific glucose uptake rate (Q_G) and the actual glucose concentration in the culture medium was observed (Fig. 8).

Loss of control over glucose (or other substrates) transport, a widespread phenomenon occuring in transformed cells [33, 77, 126], was responsible and the excess glucose taken up was wasted through overproduction and secretion of lactate during the course of the batch culture. It has also been observed that, with this particular cell cultivation procedure, the non-growth-associated utilization of glucose can be reduced and the cell yields can be improved when the addition of glucose is automatically controlled by fed-batch methods.

Increased cell yields resulting from the continuous method of cultivation, have also been observed with other cell lines (BHK 21) on serum-rich culture media (KATINGER et al., unpublished results). However, in these studies, the growth limiting substrate has yet to be identified. All the above investigations indicate that the use of continuous cultivation methods leads to improved cell yields in comparison to other cultivation procedures This may therefore be a phenomenon general to the cultivation of all animal cells.

Different principles may be responsible for the better growth results obtained by the use of continuous systems; reduced inactivation of physically unstable substance due to their shorter residence times in the continuous system as compared to batch systems, and/or the more economic use of some growth regulating factors (such as hormones) when they are continuously fed to the culture. Hormones, in particular, are destroyed by the cells as they are used [36].

Specific endogenous inhibitors of cell growth, two of which have recently been identified [36a], may be responsible for the cell density-dependent regulation of growth exhibited by some cells. Dilution of such inhibitors by fed-batch, perfusion, dialysis or by continuous culture may thus lead to improved growth in these cases.

To summarize, therefore, the controlled feeding of unstable substances, or the continuous dilution and/or removal of inhibitors, maintains their concentrations at acceptable levels. An improved efficiency in cell growth is thus effected.

From this state of affairs it can be generally concluded that it is, in principle, possible to improve the efficiency of a cell culture process several fold by simply changing the method of cultivation from a closed to an open system.

Figure 9 demonstrates one example of the way in which these principles could be employed to achieve a much more efficient production of interferon from lymphoblastoid cells. Method I of figure 9 shows the production process using batch methods in which undesirable alterations of cell phenotype are hopefully prevented by the use of nitrogen stored stock cultures as fresh inocula for each fermentation run. This is the most commonly applied practice to date. [64]. In the semi-continuous approach (II of Fig. 9), a fresh inoculum is employed only 3 to 4 times annually. The possibility of strain degeneration is therefore increased compared to the first method and additional adverse effects on the production may also occur due to altered environmental conditions during the course of prolonged cultivation. The scheme proposed by us (III of Fig. 9) uses a two-stage culture system in which homogeneous growth is continuously produced in the first stage and provides the cells with a uniform and well-developed history. The cells resulting from this procedure are continuously fed into either of two second stages, in which media has

been previously added to ensure continued cell growth. Once the desired volume is achieved in one of the second stages, the cells can be synchronized by different means [121a] prior to virus infection, and the first stage culture is fed into the alternative second stage to repeat the production cycle.
One advantage of the latter scheme is that a small first stage (growth vessel) can be operated at high growth rates and cell yields thus producing more cells per unit volume of fermenter and media. Media costs are thus substantially reduced, as are equipment and labour costs due to the continuous (first stage) and semicontinuous (second stage) modes of operation. Additionally, it is possible to use serum-free media in the second stage, since a few growth doublings can be sufficiently supported by the residual serum components supplied from the first stage.
The general applicability of these principles, however, must be checked from case to case, since unexpected results may occur at any time. A typical example of this is given in

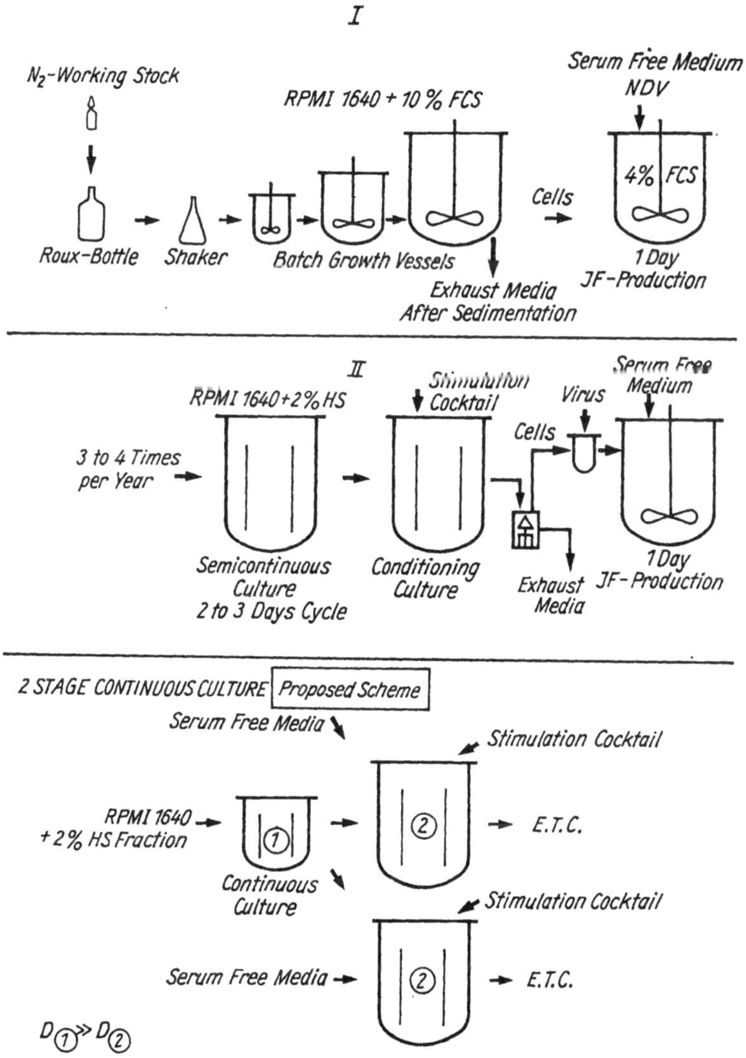

Fig. 9. Simplified schemes of production of lymphoblastoid interferon.

Figure 10, in which changes in the various quantities of acid phosphate iso-enzymes occurred in response to the method of cultivation.

These iso-enzymes, used as markers for unpredictable physiological changes during extended cultivations by different methods, provide us with a clear indication of the importance of the choice of experimental design in relation to the particular requirements of the process. Therefore, one adverse effect of the use of continuous systems, is that, whilst achieving higher cell densities, they may adversely affect the overall process by bringing about undesirable physiological changes. Further generalizations regarding the choice of cultivation method are not possible due to the specificity of individual processes.

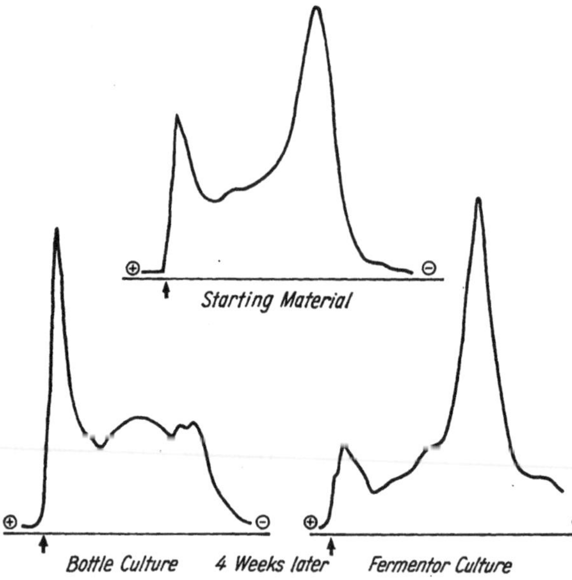

Fig. 10. Densitometer traces from polyacrylamidgels of cellular isoenzymes of acid phosphatase.
With a human lymphoblastoid leukemia cell line (REH 6, 33. passage) a airlift fermentor is inoculated and in addition, with the same media and inoculation material, a normal bottle cultivation is contiued. Electrophoretic analysis of starting material and cells sampled after 4 weeks of cultivation is performed. For technical details see [107].

6. Conclusions

In this review we have tried to discuss, on a broad basis, the current status, potential developments and possible alternatives to animal cell technology. In our opinion, recombinant DNA techniques are the most promising alternative for the production of animal cell culture derived biologicals. This is particularly the case because advances in the field of animal cell cultivation have been hindered by restrictions in the use of non-normal cells. This philosophy, which only permits the use of normal cells for most applications, has resulted in a cascade of effects detrimental to the development of a competitively economic, simpler and more reliable animal cell technology. Therefore, as long as cells are kept in a "normal" state, they cannot be accessible to genetic manipulation for process improvement, and so will require more expensive serum and other constituents in their

medium design. In addition, their quality control will remain more complex and labour intensive and the techniques necessary for their mass cultivation will remain technologically less efficient. Furthermore, current knowledge is, even today, often not applied to existing technologies for their improvement. Provided this situation prevails, animal cell technology will remain restricted to only a few specialized cases. For example, monoclonal antibodies of predetermined specificity produced by hybridomas are currently out with the scope of genetic engineering due to the diverse distribution of the necessary genetic loci over many chromosomes making attempts to transfer the required genetic information to microbial systems unrealistic. This also applies to those biologicals which require co- or post-translational processing, however, simpler proteins, such as interferon, are competitively better candidates for such genetic manipulations.
In this article we have also presented ideas, both old and new, on purely biological and engineering levels which may be used to improve the design of animal cell technologies by some orders of magnitude. We have established that there now exist much improved and more selective purification methods for proteins, cheaper formulations for cultivation media, and more economic and reliable ways of large scale animal cell cultivation. Altogether, these new developments give us cause for becoming more optimistic about the future of animal cell technologies and must surely motivate us to go ahead.

Acknowledgements:

The authors express their thanks for typing the manuscript to Mrs. Linda Bliem, Mrs. Monika Kantner and Mrs. Reinhild Nürnberger and for the corrections to Dr. Fiona Hamilton and Dr. David E. Pitt.
This work was supported in part by the „Fonds zur Förderung der wissenschaftlichen Forschung Österreichs Projekt Nr. 3562".

Eingegangen: 5. 2. 1981

References

[1] Accolla, R. S., Carrel, S., and Mach. J. P.: "Monoclonal antibodies specific for carcino embryonic antigen and produced by two hybrid cell lines." Proc. Nat. Acad. Sci. (USA) **77** (1980) 563.
[2] Acton, R. T., Barstad, P. A., Cox, R. M., Zwerner, R. K., Wise, K. S., and Lynn, J. D. (1975). "Large scale production of murine lymphoblastoid cell lines expressing differentiation alloantigens." The First Internat. Cell Culture Congress, Birmingham, Alabama. In: R. T. Acton and J. D. Lynn (eds.) "Cell culture and its application." Academic Press (1977), p. 129—161.
[3] Ashwood-Smith, M. J., and Friedmann, G. B.: "Lethal and chromosomal effects of freezing, thawing, storage time and x-irradiation on mammalian cells preserved at —196 °C in DMSO." Cryobiology, **16** (1979) 132.
[4] Barile, M. F., and Kern, J.: "Isolation of mycoplasma arginini from commercial bovine sera and its implication in contaminated cell cultures." Pros. Soc. Exp. Biol. (N. Y., USA) **138** (1971) 432.
[5] Barnes, D., and Sato, G.: Review. "Methods for growth of cultured cells in serum-free medium." Anal. Bioch. **102** (1980) 255.
[6] Barteling, S. J.: "Some aspects of FMDV-production in growing cells and a closed system for concentration of FMDV by polyethylene glycol." 2nd ESACT-Meeting, Paris, in: Dev. Biol. Stand **42** (1978) 71. S. Karger, Basel 1979.
[7] Bauer, R. F., Arthur, L. O., and Fine, R. L.: "Propagation of mouse mammary tumor cell-lines and production of mouse mammary tumor virus in a serum-free medium." In Vitro, **12**, (8), (1976) 558.
[8] "Biocarrier". Bio Rad Laboratories, Richmond, California, (USA).

[9] BIRCH, J. R.: "The role of serum in the culture of a human lymphoblastoid cell line." 3rd ESACT Meeting, Oxford, In: Dev. Biol. Stand. **40** (1979) 21. S. KARGER, Basel, 1980.
[10] BISWAS, C., and DAYER, J. M.: "Stimulation of collagenase production by collagen in mammalian cell cultures." Cell **18** (1979) 1035.
[11] BLAKEBROUGH, N., and SAMBAMURTHY, K.: "Mass transfer and mixing rates in fermentation vessels". Biotechn. & Bioeng. **8** (1966) 25.
[12] BÖTTCHER, I., and HÄMMERLING, G.: "Continuous production of monoclonal mouse IgE antibodies with known allergenic specificity by a hybrid cell line." Nature **275** (1978) 761.
[13] BOTTENSTEIN, J., HYASHI, I., and HUTCHINGS, MASUI, H., MATHER, J., MCCLURE, D. B., OHASA, S., RIZZINO, A., SATO, G., SERRERO, G., WOLFE, R., and WU, R. (1979). "The growth of cells in serum-free hormone-supplemented media." In: W. B. JACOBY, I. H. PASTAN (eds.) Methods in Enzymology LVIII, p. 94—109, Academic Press.
[14] BUSH, H., and SHODELL, M. S.: "Cell cycle changes in transformed cells growing under serum-free conditions." J. Cell. Phys. **90** (1976) 573.
[15] CHARRON, D. J., and MCDEVITT, H. D.: "Analysis of HLA-D region-associated molecules with monoclonal antibody." Proc. Nat. Acad. Sci. (USA) **76** (1979) 6567.
[16] CHEN, T. R.: "In situ detection of mycoplasma contamination in cell cultures by fluorescent Hoechst 33258 stain." Exp. Cell. Res. **104** (1977) 255.
[16a] COLBERG-GARAPIN, Florence., HORODNICEANU, F., KOURILSKY, Ph., and GARAPIN, A. C.: "Cloning of the Herpes Simplex Virus Typ I thymidine Kinase gene in *E. Coli* K12: A selective marker for gene transfer into animal cells." 3rd ESACT Meeting, Oxford, In: Dev. Biol. Stand. **40** (1979) 75. S. KARGER, Basel, 1980.
[17] CRAM, L. S., and LEHMAN, J. M.: "Flow micro fluorometric DNA content measurements of TC cells and peripheral lymphocyte." Human Genetics Vol. **37** (2), (1977) 201.
[18] CREA, R., and KRASZEWSKI, A., HIROSE, T., and ITAKURA, K.: "Chemical synthesis of genes for human insulin." Proc. Nat. Acad. Sci. (USA) **75** (1978) 5765.
[19] DENT, P. B., CLELAND, G. B., and LIAO, S. K.: "Detection and control of occult mycoplasma contamination in human tumor cell lines." Cancer Immunol. Immunther. **8** (1980) 27.
[20] Direction of the Institute of Immunology, Zagreb (eds.) "Karyologic determinations". In: Proc. on the Symposium on the Characterization and Uses of Human Diploid Cell Strains. Opatija (1963).
[21] ESPMARK, J. A., ALQUIST-ROTH, L., SÄRNE, L., and PERSSON, A.: "Tissue typing of cells in culture III. HLA antigens of established human cell lines. Attempts of typing by the mixed HA technique." Tissue Antigens **11** (1978) 279.
[21a] FAZEKAS de St. GROTH, S., and SCHEIDEGGER, D.: "Production of monoclonal antibodies: Strategy and tactics". J. Immunol. Methods **35** (1980) 1.
[22] HONMA, Y., KASUKABE, T., OKABE, J., and HOZUMI, M.: "Replacement of serum by insulin transferrin, albumin, phosphatidyl choline, cholesterol, and some trace elements in cultures of mouse myeloid leukemia cells sensitive to inducers of differentiation." Exp. Cell Res. **124** (1979) 421.
[23] FINTER, N. B., and BRIDGEN, P. J. (1978). "The large scale production of human interferons and their possible uses in medicine." TIBS — April 1978.
[24] Flow Laboratories, Irvine, Scotland,
"Superbeads."
[25] GALFRE, G., HOWE, S. C., MILSTEIN, C., BUTCHER, G. W., and HOWARD, J. C.: "Antibodies to major histocompatibility antigens produced by hybrid cell-lines." Nature, **256** (1977) 550.
[26] GATTI, R. A., and LEIBOLD, W.: "HLA-D typing with lymphoblastoid cell-lines VI and VII allelic relationships". Tissue Antigens **14** (1979) 183.
[27] DI GIACOMO, F. R., and SHAH K. V.: "Virtual absence of infection with Herpesvirus simiae in colony reared rhesus monkeys (Macaca mulatta), with a literature review on antibody prevalence in natural and laboratory rhesus populations." Laboratory Animal Science, **22** (1972) 61.
[28] GIRARD, H. G., OKAY, G., and KIVILCIM, Y.: "Use of the vibro fermenter for multiplication of BHK cells in suspension and for replication of FMD virus." Bull. off. Int. Epiz. **79** (1973) 805.
[29] GIRARD, H. C. (1975). "Problems encountered in large scale cell production plants". The First Intern. Cell Culture Congress, 1975, Birmingham, Alabama. In: R. T. ACTON and J. D. LYNN (eds.) "Cell culture and its application". Academic Press (1977).

[30] GIRARD, H. C., SÜTCÜ, M., ERDOM, H., and GÜRHAM, I.: "Monolayer culture of animal cells with the gyrogen equipped with tubes". 2nd ESACT-Meeting, Paris 1978. In: Develop. Biol. Stand. **42** (1978) 127, S. KARGER, Basel 1979.
[31] HAM, R. G., and SULLIVAN, P. M. (1975). "Qualitative and quantitative analysis of nutritional factors for cell growth." First International Cell Culture Congress, Birmingham, Alabama, USA. In: R. T. ACTON, and D. J. LYNN (eds.) "Cell Culture and its application". Academic Press (1977), p. 533–541.
[32] HAM, R. G., and MCKEEHAN, W. L. (1979). "Media and growth requirements" in methods in enzymology Vol. LVIII. W. B. JACOBY and I. H. PASTAN, (eds.). Academic Press (N. Y.).
[33] HATANAKA, M., and HANAFUSA, H.: "Analysis of a functional change in membrane in the process of cell transformation by rous sarcoma virus; alteration in the characteristics of sugar transport." Virology **41** (1970) 647.
[34] HEALY, G. M., and MACMORINE. H. G.: "Replacement of serum by a soy bean protein fraction in a basal tissue culture medium." Progr. Immunol. Standard, **5** (1972) 202.
[35] HIGUCHI, K.: "Cultivation of animal cells in chemically defined media." A review. Adv. Appl. Microbiol. **16** (1973) 111.
[36] HOLLEY, R. W.: "Control of growth of mammalian cells in cell culture." Nature **258** (1975) 487.
[36a] HOLLEY, R. W., BÖHLEN, P., FAVA, R., BALDWIN, J. H., KLEEMAN, G., and ARMOUR, R.: "Purification of kidney epithelial cell growth inhibitors." Proc. Natl. Acad. Sci. USA **77** (1980) 5989.
[37] HONG, S. L., PATTON, G., and DEYKIN, D.: "Archadonic acid level in cellular lipids determines the amount of prostaglandin synthesized during cell growth in tissue culture." Prostaglandins **17** (1979) 53.
[38] HONMA, Y., KASUKABE, T., OKABE, J., and HOZUMI, M.: "Replacement of serum by insulin, transferrin, albumin, phosphatidyl choline, cholesterol and some trance elements in cultures of mouse myeloid leukemia cells sensitive to inducers of differentiation." Exp. Cell. Res. **124** (1979) 421.
[39] HUTCHINGS, S. E., and SATO, G. H.: "Growth and maintenance of HeLa cells in serum-free medium supplemented with hormones." P.N.A.S. (USA) **75** (1978) 901.
[40] HUSSA, R. D.: "Latent ability to produce human chorionic gonadotropin in response to dibutyryl adenosine 3'5'-monophosphate in the BeWo-NBC tropoblastic cell line". Journ. Clin. Endocrin. and Metabolism **50** (1), (1980) 5.
[41] INGLOT, A. D., GODZINSKA, H., and CHUDZIO, T.: "Use of polyethylene glycol-treated calf serum for cell cultures in virus and interferon studies." Acta. Virol. **19** (1975) 250.
[42] ISCOVE, N. N., and MELCHERS, F.: "Complete replacement of serum by albumin, transferrin, and soy bean lipid in cultures of LPS-reactive B-lymphocytes." Journ. Exp. Med. **147** (1978) 923.
[43] ITAKURA, R., HIROSE, T., CREA, R., RIGGS, A. D., HEYNECKER, H. L., BOLIVAR, F., and BOYER, H. W.: "Expression in E. coli of a chemically synthesized gene for human somatostatin." Science, **198** (1977) 1056.
[44] JENKIN, H. M., YANG, T. K., and ANDERSON, L. E.: "The effect of partially hydrolyzed casein on the growth of human skin diploid cells in vitro". Proc. Soc. Exp. Biol. Med. **160** (1979) 59.
[45] JACOBS, J. P.: "Serially propagated human diploid cells: a synopsis of the present position concerning their use for producing viral vaccines and interferon." 2nd ESACT-Meeting, Paris 1978. In: Dev. Biol. Stand. **42** (1978) 13. S. KARGER, Basel 1979.
[46] JACOBS, J. P.: "A study of the longevity and other properties of human diploid fibroblasts propagated in several nutrient media." 3rd ESACT-Meeting, Oxford, 1979. In: Dev. Biol. Stand. **46** (1979) 49. S. KARGER, Basel 1980.
[47] JOHNSTON, M. G., CHRISTOFINIS, G., BALL, G. D., FANTES, K. H., and N. B. FINTER: "A culture system for producing large amounts of human lymphoblastoid interferon." 2nd ESACT-Meeting, Paris, 1978. In: Dev. Biol. Stand. **42** (1978) 189. S. KARGER, Basel 1979.
[48] JUNG-TESTAS, I., and BAULIEU, E. E.: "Effects of sex steroids and antihormons on growth, adhesiveness and receptors of L-929 cells cultured in serum-containing and serum-free media." Exp. Cell. Res. **119** (1979) 75.
[49] KATINGER, H. W. D., and FRENCL, H. (1979).
Unpublished results.

[50] KATINGER, H. W. D., SCHEIRER, W., and KRÖMER, E.: "Bubble column reactor for mass propagation of animal cells in suspension culture." German Chemical Engineering, **2** (1979) 31.
[51] KATINGER, H. W. D. (1977). "New fermenter configurations in biotechnology and fungal differentiation." J. MEYRATH, and Bullock (eds.). Academic Press 1977.
[52] KATSUTA HAJIM (eds.) (1978). "Nutritional requirements of cultured cells." Japan Scientific Societies Press Tokyo, University Park Press, Baltimore.
[53] KEAY, L. (1975). "Autoclavable low cost cell culture media." The First Internat. Cell Culture Congress, Birmingham, Alabama, USA. In: R. T. ACTON, and J. D. LYNN (eds.). "Cell culture and its application". Academic Press (1977), 513.
[54] KEAY, L.: "Autoclavable low cost serum free cell culture media: the growth of established cell-lines and production of viruses." Biotechn. Bioeng. Vol. 18 (1976) 363.
[55] KELLER, W.: "Comments on tissue growth factors." Vox sanguinis, **3** (1958) 52.
[56] KENNETT, R. H., GILBERT, F.: "Hybrid myelomas producing antibodies against a human neuroblastoma antigen present on fetal brain." Science, **203** (1979) 1120.
[57] KENNETT, R. H.: "Monoclonal antibodies hybrid myeloma — a revolution in serology and immunogenetics." Am. J. Hum. Genet. **31** (1979) 539.
[58] KILBURN, D. G., and WEBB, F. C.: "The cultivation of animal cells at controlled dissolved oxygen partial pressure." Biotechn. Bioeng. **X** (1968) 801.
[59] DE KINKELIN, P., and LE BERRE, M.: "Mass virus production in fish-cell system." 2nd ESACT-Meeting, Paris, 1978. In: Dev. Biol. Stand. **42** (1978) 99. S. KARGER, 1979.
[60] KLEIN, F., ROSENSTEEL, J. F., and HUMMER, R. M., et al.: "Large-scale production and concentration of infectious Eppstein-Barr-Virus." Appl. Environm. Microbiol. **35** (1978) 172.
[61] KLEIN, F., RICKETTS, R., JONER, W. I., DE ARMON, I. R., TEMPLE, M. J., ZOON, K. C., and BRIDGEN, P. J.: "Large-scale production and concentration of human lymphoid interferon." Antimicr. Agents and Chemother. **15** (1979) 420.
[61a] KÖHLER, G., MILSTEIN, C.: "Continous cultures of fused cells secreting antibody of predefined specifity." Nature **256** (1978) 495.
[62] KONOWALCHUK, J., and SPEIRS, J. I.: "Response of various cell lines to E. coli toxic products." Can. J. Microbiol. **25** (1979) 338.
[63] LAMBERT, K., and PIRT, S. J.: "Growth of human diploid cells (strain MRC-5) in defined medium: Replacement of serum by a fraction of serum ultrafiltrate." J. Cell. Sci. **35** (1979) 381.
[64] LIPSON, L. G., BEITINS, I. Z., KORNBLITH, P. L., MCARTHUR, J. W., FRIESEN, H. G., KLIMAN, B., KJELLBERG, R. N.: "Tissue culture studies on human pitutary tumors: long term release of anterior pitutary hormons into the culture medium." Acta Endocrim. **90** (1979) 421.
[65] LITTLEFIELD, J. W.: "Selection of hybrids from matings of fibroblasts in vitro and their presumed recombinants." Science **145** (1964) 709.
[66] MACKOWIAK, C., REGAMEY, R. H. eds.: "Developments in biological standardization, Vol. 35." Food and Mouth Disease Vaccines, Lyon, 1976. S. KARGER, Basel, 1977.
[67] MCCLANE, B. A., and MCDONEL, J. L.: "The effects of clostridium perfrigens enterotoxinon morphology, viability and macromolecular synthesis in VERO-cells." J. Cell. Phys. **99** (1979) 191.
[68] MCGARRITY, G. J., VANAMAN, V., and SARAMA, J. (1978). "Mycoplasma infection of cell cultures." G. J. MCGARRITY, D. G. MURPHY, W. W. NICHOLS (eds.) p. 213—214. Plenum, New York.
[69] MCKEEHAN, W. L., MCKEEHAN, K. A., HAMMOND, S. L., and HAM, R. G.: "Improved medium for clonal growth of human diploid fibroplasts of low concentrations of serum protein." In Vitro **13**, No. 7 (1977) 399.
[70] MCKEEHAN, W. L., GENEREUX, D. P., and HAM, R. G.: "Assay and partial purification of factors from serum that control multiplication of human diploid fibroplasts." Biochem. Bioph. Res. Comm. **80** (1978) 1013.
[71] MCKEEHAN, W. L., and MCKEEHAN, K. A.: "Oxocarboxylic acids, pyridine nucleotide linked oxidoreductases and serum factors in regulation of cell proliferation." J. Cell. Phys. **101** (1979) 9.
[72] MCLIMANS, W. F. (1979). "Mass culture of mammalian cells." In: W. B. JACOBY, I. H. ASTAN (eds.) "Methods in enzymology" LVIII, p. 194—211. Academic Press, 1979.

[73] MEIGNIER, B., MOUGEOT, H., and FAVRE, H.: "Foot and mouth disease virus production on micro carrier grown cells." 3rd ESACT-Meeting, Oxford. In: Dev. Biol. Stand. **46** (1979) 249. S. KARGER, Basel, 1980.
[74] MELCHERS, F., POTTER, M., WARNER, N. L. (eds.) (1978). Current topics in microbiology and immunology Vol. 81." Lymphocyte hybridomas". Springer Verlag, 1978.
[74a] MILSTEIN, C. (1980). "Monoclonal Antibodies." Scientific American Sept. 1980, 56–62.
[75] MILTENBURGER, H. G.: "The replication of insect pathogenic viruses in cell culture". 3rd ESACT-Meeting, Oxford, 1979. In: Dev. Biol. Stand. **46** (1980) 295.
[76] MIZRAHI, A., and SHAHAR, A.: "Partial replacement of serum by vegetable proteins in BHK—culture medium." Journ. Biol. Stand. **5** (1977) 327.
[77] MOOLTEN, F. L., MOOLTEN, D. N., and CAPPARELL, N. J.: "Transformed BHK cells exhibiting normal or subnormal sugar uptake." J. Cell. Physiol. **93** (1977) 147.
[78] MOORE, G. E., HASENPUSCH, P., GERNER, R. E., and BURNS, A. A.: "A pilot plant for mammalian cell culture." Biotechn. Bioeng. **X** (1968) 625.
[79] MOOREHEAD, P. (Chairman).: "Report of ad hoc comittee on karyological controls of human cell substrates." Journ. Biol. Stand. **7** (1979) 397.
[80] MORRISSEY, J. J., HAMILTON, J. W., and COHN, D. V.: "The secretion of parathormone and glycosilated proteins by parathyroid cells in culture." Bioch. Bioph. Res. Comm. **82** (1978) 1279.
[81] MOSKOWITZ, M., CHENG, D. K. S., MOSKATELLO, D. K., and OTSUKA, M.: "Growth stimulation of BHK cells in culture by free biotin in serum." Jnl. Nat. Cancer Inst. **64**, (1980) 639.
[82] MOWAT, G. N., GARLAND, A. J., and SPIER, R. E.: "The development of FMD vaccines." The Veterinary Record **102** (9), (1978) 190.
[82a] MULLIGAN, R. C., and BERG, P.: "Expression of a bacterial gene in mammalian cells." Science **209** (1980) 1422.
[82b] NADLER, L. M., STASHENKO, Ph., HARDY, R., KAPLAN, W. D., BUTTON, L. N., KUFE, D. W., ANTMANK, H., and SCHLOSSMAN, S. F.: "Serotherapy of a patient with monoclonal antibody directed against a human lymphoma associated antigen". Cancer Research. **40** (1980) 3147.
[83] NAGATA, S., TAIRA, H., HALL, A., JOHNSRUD, L., STREULI, M., ECSÖDLI, J., BOLL, W., CANTELL, C., and WEISSMANN, C.: "Synthesis in E. coli of a polypeptide with human leucocyte interferon activity." Nature **284** (1980) 316.
[84] NAGLE, S. C., and BROWN, B. L.: "An improved heatstable glutamine-free chemically defined medium for growth of mammalian cells." J. Cell. Phys. **77** (1971) 259.
[85] NAGLE, S. C., and BROWN, B. L.: "Epstein Barr Virus from P3HR-1 cells grown in chemically defined medium." Applied Microbiol. **28** (1974) 518.
[86] NAGLE, S. C., FINE, D. L.: "Demonstration of components of serum-free culture medium effecting maximum in vitro expression of mouse mammary tumor virus." In Vitro **14** (1978) 218.
[87] NILAUSEN, K.: "Role of fatty acids in growtpromoting effect of serum albumin on hamster cells in vitro." J. Cell. Phys. **96**, (1), (1978) 1.
[88] A/S Nunc, Kamstrup, Denmark. "Biosilon".
[89] NYIRI, L. N. (1975). "Computer control of mammalian cell suspension cultures". The First Internat. Cell Culture Congress, 1975. In: R. T. ACTON, J. D. LYNN (eds.) "Cell culture and its application". Academic Press (1977), 161–191.
[90] OLDSHUE, J. Y.: "Fermentation mixing scale up techniques". Biotechn. Bioeng. **8** (1966) 3.
[91] OLDSHUE, J. Y.: "Fluid mixing". Process Biochem. **4** (1969) 51.
[91a] OLSSON, L., and KAPLAN, H. S.: "Human-human hybridomas producing monoclonal antibodies of predefined antigenic specificity." Proc. Natl. Acad. Sci. U.S.A. **77** (1980) 5429.
[92] PAPERMASTER, B. W., MCENTRE, J. E., and SKISHAK, C. M. (1978). "Immunotherapy with lymphokine preparations". In: B. SERROU, C. ROSENFELD (eds.) "Human Lymphocyte Differentiation". Inserm. Symp. No. 8. Elesevier/North Holland, 1978.
[93] PERKINS, F. T., and REGAMEY, R. H. eds. (1976). "Developments in biological standardization Vol. 37." Standardization of cell substrates for the production of virus vaccines, Geneva, 1976. S. KARGER, Basel 1977.
[94] PETRICCIANI, J. C., HOPPS, HOPE E., and CHAPPLE, Paul J. (1979). "Cell substrates, their use in the production of vaccines and other biologicals." Adv. In. Exp. Med. and Biol. Vol. 118. Plenum Press, New York, 1979.

[95] Pharmacia Fine Chemicals AB., Uppsala, Sweden, "Cytotex 1".
[96] Pickart, L., and Thaler, M. M.: "Tripeptide in human serum which prolongs survival of normal liver cells and stimulated growth in neoplastic liver." Nature **243** (1973) 85.
[97] Plotkin, S. A., and Huygelen, G.: "Cytomegalovirus vaccine prepared in Wi-38." In: Dev. Biol. Stand. **37** (1976) 301. S. Karger, Basel.
[98] Poiley, J. A., Schuman, R. F., and Pienta, R. J.: "Characterization of normal human embryo cells grown to over 100 population doublings." In Vitro **14** (1978) 405.
[99] Pollard, R., and Khosrovi, B. (1978). "Reactor design for fermentation of fragile tissue cells." Process Biochem. May. 1978, p. 31–37.
[100] Pye, E. K. (1977). "Production of enzymes in tissue culture". In: Z. Bohak, and N. Sharon (eds.). "Biotechn. applications of proteins and enzymes." Academic Press, 1977.
[101] Raschke, W. C.: "Plasmacytomas, lymphomas and hybridomas: their contribution to immunology and molecular biology." Biochem. Biophys. Acta **605** (1980) 113.
[102] Rippin, D. W. T.: "The recycle reactor as a model of incomplete mixing". Ind. and Eng. Chem. Fundamentals **6** (1967) 488.
[103] Ritz, J., Pesando, J. M., Notis-McConarty, J., Lazarus, H., and Schlossman, S. F.: "A monoclonal antibody to human acute lymphoblastic leukemia antigen." Nature **283** (1980) 583.
[104] Rosenfeld, C., Martyre, M. C., and Pico, J. L. (1978). "Use of a human cell line in acute lymphoid leucemia immunotherapy." Inserm Symposium No. 8. In: B. Serrou and C. Rosenfeld (eds.) "Human Lymphocyte Differentiation". Elsevier/North Holland, 1978.
[105] Schaeffer, W. I.: "Proposed usage of animal tissue culture terms", revised 1978. In Vitro **15** (9), (1978) 649.
[106] Scheirer, W., Katinger, H., Laber, G.: "Experience with testing for mycoplasma with five different methods." 2nd ESACT-Meeting, Paris, 1978. In: Dev. Biol. Stand. **42** (1978) 55. S. Karger, Basel 1979.
[107] Scheirer, W., Nowotny, H., and Lutz. D.: "Alteration of a human lymphoblastoid cell line. REH6 during subsequent cultivation." 3rd ESACT-Meeting, Oxford 1979. In: Dev. Biol. Stand. **46** (1979) 101. S. Karger, Basel, 1980.
[108] Secher, D. S., and Burke, D. C.: "A Monoclonal antibody for large-scale purification of human leukocyte interferon." Nature **285** (1980) 446.
[109] Sethi, K. K., and Brandis, H.: "Production of neutralizing antibody to rabies and VSV by hybrid cell lines." Cellular Immunology **49** (1980) 260.
[110] Spier, R. E.: "Advantages of a microprocessor-monitored and controlled continuous culture of BHK-suspension cells." 3rd ESACT-Meeting, 1979. In: Dev. Biol. Stand. **46** (1979) 159. S. Karger, Basel, 1980.
[111] Spier, R. E.: "Recent developments in the large scale cultivation in animal cells in monolayers." Advances in Biochemical Engineering **14** (1980) 119.
[112] Van Steenis, G., De Groot, J. G. M., Van Rossum, G. W. A., and Kruijt, B. C. (1979). "Virus infection in captive-bred versus wild-caught cynomolgus monkeys." Proceedings VIIIth ICLA-Symposium on animal quality and models in biomedical research. Utrecht, August 1979. Gustav Fisher Verlag, Stuttgart/New York. In press.
[113] Steinitz, M., Sepälä, I., Eichmann, K., and Klein, G.: "Establishment of a human lymphoblastoid cell line with specific antibody production against group A staphylococcal carbohydrate." Immunobiol. **156** (1979) 41.
[114] Strong, D. M., Pellegrino, M. E., and Perrone, S.: "Human lymphoid cell lines as targets for DRW." In Vitro **14** (1978) 989.
[115] Tashijan, A. H. (1979). "Clonal strains of hormone-producing pitutary cells." In: W. B. Jacoby and I. H. Pastan (eds.) "Methods in Enzymology" Vol. LVIII (1979) p. 527–536. Academic Press, 1979.
[116] Taub, M., Chuman, L., Saier, M. H., and Sato, G.: "Growth of MDCK cells in hormone-supplemented, serum-free medium." Proc. Nat. Acad. Sci (USA) **76** (1979) 3338.
[117] Telling, R. C., and Radlett, P. J.: "Large scale cultivation of mammalian cells". Advances in Appl. Microbiol. **13** (1971) 91.
[118] Telling, R. C., and Stone, C. J.: "A method of automatic pH control of a bicarbonate-CO_2 buffer system for the submerged culture of hamster kidney cells." Biotechn. Bioeng. **VI** (1964) 147.

[119] TODARO, G. J.: "Rapid detection of mycoplasma-infected cell cultures". Exp. Cell Res. **65** (1971) 256.
[120] TOMEI, L. D., and C. J. ISSEL: "Growth and immunogenenicity fo FMD-virus in BHK-cells adapted to and continuously grown in a serum free chemically defined média." Biotechn. Bioeng. XVII (1975) 765.
[121] TOVEY, M. G., and BROUTY-BOYÉ, D.: "Characteristics of the chemostat culture of murine leukemia L1210 cells". Exp. Cell Res. **101** (1976) 346.
[121a] TOVEY, M. G., BEGON-LOURS, J., GRESSER. T. and MORRIS, A. G.: "Marked enhancement of Interferon production in 5-Bromodeoxyuridine treated human lymphoblastoid cells. Nature, **267** (1977) 455.
[122] TSAI, B., ERICKSON, L. E., and FAN, L. T.: "The effect of micromixing and growth process". Biotechn. Bioeng. **11** (1969) 181.
[123] TSUCHIYA, S., YOKOYAMA, S., YOSHIE, O., and ONO, Y.: "Production of diphteria antitoxin antibody in Epstein-Barr-Virus-induced lymphoblastoid cell lines." Journ. of Immunol. **124** (1980) 1970.
[124] WANG, D. I. C. (1979). Personal communication.
[125] WAYMOUTH, C. (1977). "Studies on chemically defined média and the nutritional requirements of cultures of epithelial cells." Int. Symp. "Nutritional requirements of mammalian cells in culture", Tokyo, 1977. In: H. KATSUTA (ed.). "Nutritional Requirements of Cultured Cells". Japan Scientific Societies Press, Tokyo, 1978.
[126] WEBER, M. J.: "Hexose transport in normal and rous sarcoma virus-transformed cells." J. Biol. Chem. **248** (9), (1973) 2978.
[127] WERB, Z., and AGGELER, J.: "Proteases induce secretion of collagenase and plasminogen anctivator by fibroblasts." Proc. Nat. Acad. Sci. (USA), **75** (1980) 1839.
[128] VAN WEZEL, A. L.: "Growth of cell strains and primary cells on micro-carriers in homogenous culture." Nature **216** (1967) 64.
[129] VAN WEZEL, A. L., VAN DER VELDEN DE GROOT, LA. M., and VAN HER WAARDEN, J. A. M.: "The production of inactivated polio vaccine on serially cultivated kidney cells from captive bred monkeys." 3rd ESACT-Meeting, Oxford, 1979. In: Dev. Biol. Stand. **46** (1979) 151. S. KARGER, Basel 1980.
[130] WHIPPLE, A. P., DALUIN, M., and MILLS, A. J. T.: "Serum-independent human lymphoblastoids cells". Exp. Cell Res. **116** (1978) 457.
[131] WIDMANN, L. E., GOLDEN, J. J., and Chasin, L. A.: "Immortalization of normal liver functions in cell culture. Rat hepatocyte-hepatoma cell hybrids expressing ornithine carbarmyl transferase activity." J. Cell Phys. **100** (1979) 391.
[132] WILSON, L. L., and DOWDLE, W. R.: "Secretion of plasminogen activator by normal reactive and neoplastic human tissues cultured in vitro." Int. J. Cancer **22** (1978) 390.
[133] WORTHINGTON, "Limulus amebocyte lysate pyrostat™." Millipore Corporation, FREEHOLD, N. J., USA.
[134] YAMANE, I. (1978). "Development and application of a serum-free culture medium for primary culture." In: H. KATSUTA (eds.) p. 1—23. "Nutritional Requirements of Cultured Cells." Japan Scientific Societies Press Tokyo, 1978, University Park Press, Baltimore, USA.
[135] ZURAWSKI, V. R., HABER, E., and BLACK, P. H.: "Production of antibody to tetanus toxoid by continuous human lymphoblastoid cell lines." Science, **199** (1978) 1439.
[136] ZWERNER, R. K., RUNYAN, C., COX, R. M., LYNN, D., and ACTON, R. T.: "An evaluation of suspension culture systems for the growth of murine lymphoblastoid lines expressing TL and Thy-1 alloantigens." Biotechn. Bioeng. XVII (1975) 629.

Acta Biotechnologica 2 (1982) 1, 42

BOOK REVIEW

BEKER, M. E.

„Biotechnology of microbial synthesis"

Riga 1980, Sinatne publishing house
350 pp, 74 fig., 59 tab., 582 ref. (Russian)

The monograph analyses a wide range of research concerning microbiological production of biologically active substances. SCP-production is not subject of discussion.
Carbon and nitrogen sources as well as supplines, biofactors, precursors and other components of culture medium are detailed demonstrated. The utilization of industrial wastes is dealt with, too.
On the example of L-Lysine biosynthesis energetic metabolism, regulatory mechanisms of bacterial metabolic enzyme activity and bioenergetics of fermentation process are discussed, as is the technological scheme and equipment for aerobic sterile fermentation process with the analysis of aeration, stirring and foambreaking systems.
Methods are presented of both batch and continuous fermentation, also surface fermentation with special examples.
Anaerobic fermentation is discussed, since it is of great interest because of obtaining liquid and gaseous fuel and utilization of industrial and agricultureral wastes.
Further fundamentals of separation, concentration and purification of microbial metabolites with special emphasis of L-Lysine production process are given.
Exemplarily the book presents the scientific problems of biotechnological production of microbial metabolites. It would be increased the understanding to emphasize the bioengineering aspects some more for instance with illustrations of apparatus and plants and demonstration of technological problems.
The work is intended for specialists in the fields of biotechnology, bioengineering and microbiological industry and can help especially engeneering personnel for understanding the scientific problems of biotechnology.

W. BECHSTEDT

Acta Biotechnologica 2 (1982) 1, 43—50

Der Einfluß der Flockenbildung auf die Versorgung der Hefezellen mit Sauerstoff bei der Fermentation flüssiger Kohlenwasserstoffe

F. Glombitza

Akademie der Wissenschaften der DDR
Institut für technische Chemie Leipzig
7050 Leipzig, Permoserstr. 15

Summary

Floc formation, especially the influence of floc diameter variations on the total velocity of the process, was investigated in aerobic growth processes of yeast on the hydrocarbons of crude oil.
The experimental results show that the diameter of the flocs is a function of the rheological properties of the fluids and the flow conditions.
The floc diameter varies between 0,1 mm and a few millimeters. About 90% of the total yeast cells are situated in the interior of the flocs.
Since oxygen must be transferred to all yeast cells their oxygen supply was studied.
Thus, the yeast cells in the floc interior were not sufficiently supplied with oxygen, if the floc diameter reached a critical value.
In such cases a decrease of the biomass formation rate was observed, although the dissolved oxygen concentration of the aquaeous fermentation medium was greater than zero. Therefore, aerobic microbial growth processes in multicomponent systems must be carried out without floc formation or under such conditions as cause very small floc diameters.

Zusammenfassung

Durch Untersuchungen am Mehrphasensystem Hefe/Dieselkraftstoff/Wasser/Luft konnte die Bildung stabiler Systeme (Flocken) nachgewiesen werden.
Die Flockengröße ist von rheologischen Eigenschaften und Strömungsverhältnissen abhängig. Sie liegt zwischen 0,1 und einigen Millimetern.
Ungefähr 90% der vorhandenen Hefen befinden sich im Flockeninneren. Es wurde deshalb untersucht, inwieweit eine ausreichende Versorgung der Zellen mit Sauerstoff gesichert ist.
Ein Vergleich der aktuellen Verbrauchsgeschwindigkeit flockenfreier Systeme mit dem Diffusionskoeffizienten führt zu Stoffübergangsgeschwindigkeiten, die keine optimale Sauerstoffversorgung erwarten lassen.

Problemstellung

Aerobe Fermentationen sind Stoffwandlungsprozesse, an denen mindestens 3 Phasen beteiligt sind. Dadurch existieren eine große Anzahl Stoffübergangs- und Transportprozesse [1]. Die Reaktion mit der kleinsten Geschwindigkeit bestimmt die Gesamtgeschwindigkeit der Konsekutivreaktion.

Deshalb ist die Kenntnis der Geschwindigkeit der einzelnen Partialreaktionen für die gezielte Beeinflussung des limitierenden Schrittes zur Steigerung der Gesamtgeschwindigkeit notwendig.
Bei der Fermentation flüssiger Kohlenwasserstoffe durch Hefen wird die Bildung eines stabilen 4 Phasensystems, bestehend aus Hefezellen, Wasser, Luftblasen und Kohlenwasserstofftropfen beobachtet [2, 3, 4].
Diese Konglomerate, im folgenden Flocken genannt, besitzen unterschiedliche Größe und bewegen sich im Fermentationssystem.
Randwinkelmessungen zeigten, daß der stabilste Aufenthalt der Hefen zwischen der Kohlenwasserstoff- und der wäßrigen Phase ist [4].
Die Hefen belegen die Oberfläche der Kohlenwasserstofftropfen und verhindern weitgehend deren Koaleszenz. Die Existenz dieser Flockenstrukturen ist somit von der Grenzflächenspannungsdifferenz zwischen der wäßrigen und der Kohlenwasserstoffphase abhängig.
Bei einer Oberflächenspannung der wäßrigen Phase unterhalb von $35 \cdot 10^{-3}$ Jm^2 wird eine weitgehende Ablösung der Hefen vom Kohlenwasserstofftropfen beobachtet [1, 4].
Die Oberflächenspannungen der wäßrigen Phase liegen während der Fermentation jedoch bei Werten zwischen 40 und $50 \cdot 10^{-3}$ Jm^2.
Dadurch ist bei normalen Fermentationen die Existenz solcher Konglomerate immer zu erwarten.
Es ist deshalb notwendig zu wissen, welche Größen diese Strukturen annehmen, inwieweit sie Veränderungen unterliegen und welchen Einfluß sie auf die Reaktionsgeschwindigkeit der Zellvermehrung besitzen. Das ist besonders dann zu erwarten, wenn die Hefen im Flockeninneren nicht ausreichend mit Sauerstoff und Nährstoffen versorgt werden.

Untersuchungsmethode

Es werden Hefen der Gattung *Lodderomyces elongisporus* in einem Fermentor auf flüssigen Kohlenwasserstoffen gezüchtet. Als Kohlenwasserstoffphase wird eine Erdölfraktion mit einem Siedebereich von 240—360 °C (513—633 °K) und einem n-Alkangehalt von 18 Masse % verwendet.
Das Fermentationsmedium wird kontinuierlich aus dem Fermentor durch eine Beobachtungskammer gefördert.
Die Flockengrößen sowie ihre Veränderungen innerhalb der Kammer können mit Hilfe einer Kamera gefilmt werden. Die Aufnahmegeschwindigkeit beträgt maximal 3000 Bilder/Sekunde. Die Auswertung erfolgt durch Auszählen der Flockenanzahl, Bestimmung ihres Durchmessers und Einteilen in Größenklassen.
Neben der Größenbestimmung wurde besonderer Wert auf die Veränderungen der Flockengrößen gelegt.
Das ist deshalb interessant, da in einem Fermentor Zonen mit unterschiedlich starken Scherkräften und Turbulenzgraden zu erwarten sind, die auch zu unterschiedlichen Flockengrößen führen können. Zur Bestimmung der Flockengröße in Abhängigkeit von der Strömungsgeschwindigkeit wurden 4 verschiedene Geschwindigkeiten eingestellt und die sich dabei einstellenden Größenklassen als Durchgangs- und Rückstandssummenkurven dargestellt.
Aus den Flockengrößen wurde eine mittlere Flockengröße ermittelt. Als Maß für den Turbulenzgrad ist die REYNOLDzahl berechnet worden.
Die Veränderungen der Flockengröße wurden dadurch erhalten, daß das strömende Medium in den Ruhestand versetzt und die Vergrößerung der Flocken in Abhängigkeit von der Zeit gefilmt wurde.

Experimentelle Ergebnisse

Abbildung 1 zeigt die Rückstands- und Durchgangssummenkurve für verschiedene Strömungsgeschwindigkeiten. Aus ihnen wird eine Flockengröße zwischen 0,1 und 0,6 mm² Oberfläche sichtbar, wobei die kleineren Größenklassen zunehmend mit größerer Geschwindigkeit auftreten.

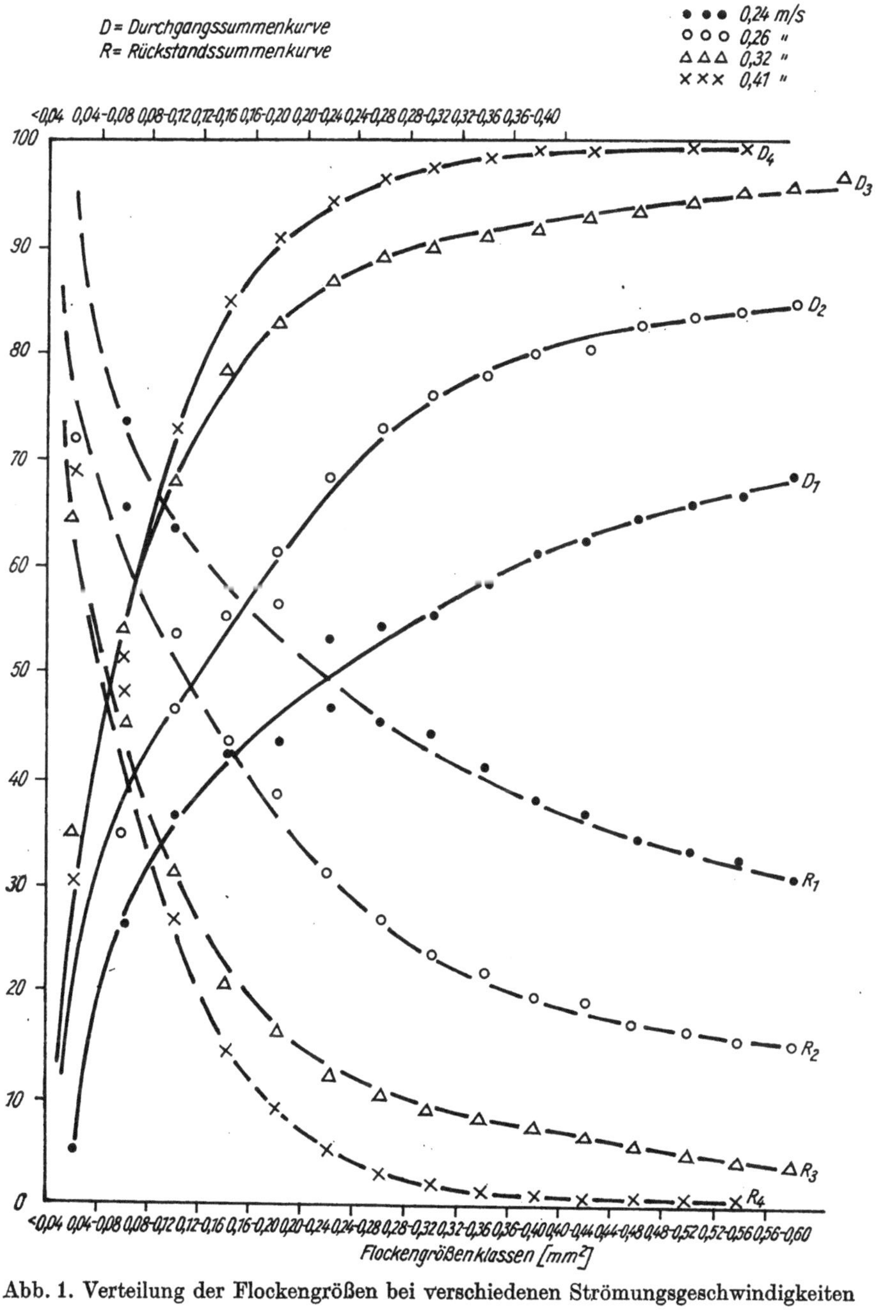

Abb. 1. Verteilung der Flockengrößen bei verschiedenen Strömungsgeschwindigkeiten

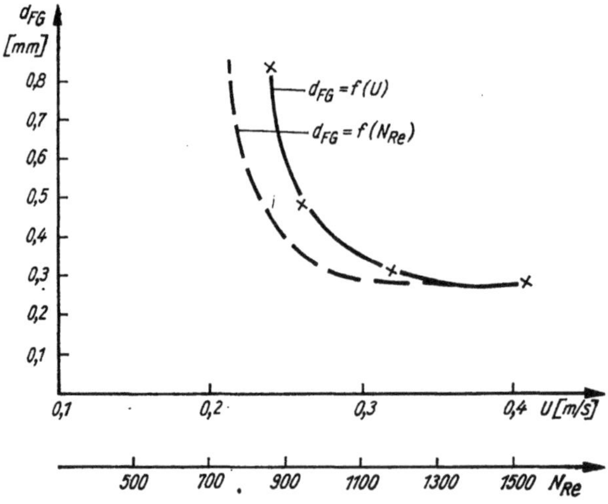

Abb. 2. Mittlere Flockengröße in Abhängigkeit von der Strömungsgeschwindigkeit und REYNOLDzahl

Das wird auch aus Abbildung 2 deutlich. Sie gibt den mittleren Durchmesser als Funktion der Strömungsgeschwindigkeit und Reynoldzahl an und zeigt selbst bei hohen Geschwindigkeiten sowie Reynoldzahlen über 1000 einen mittleren Flockendurchmesser, der nicht unter 0,3 mm liegt.

In Abbildung 3 wird das Zusammenlagern der Flocken gezeigt. Die Kurven geben den oberen und den unteren Bereich der bestimmten Flockengrößen an.

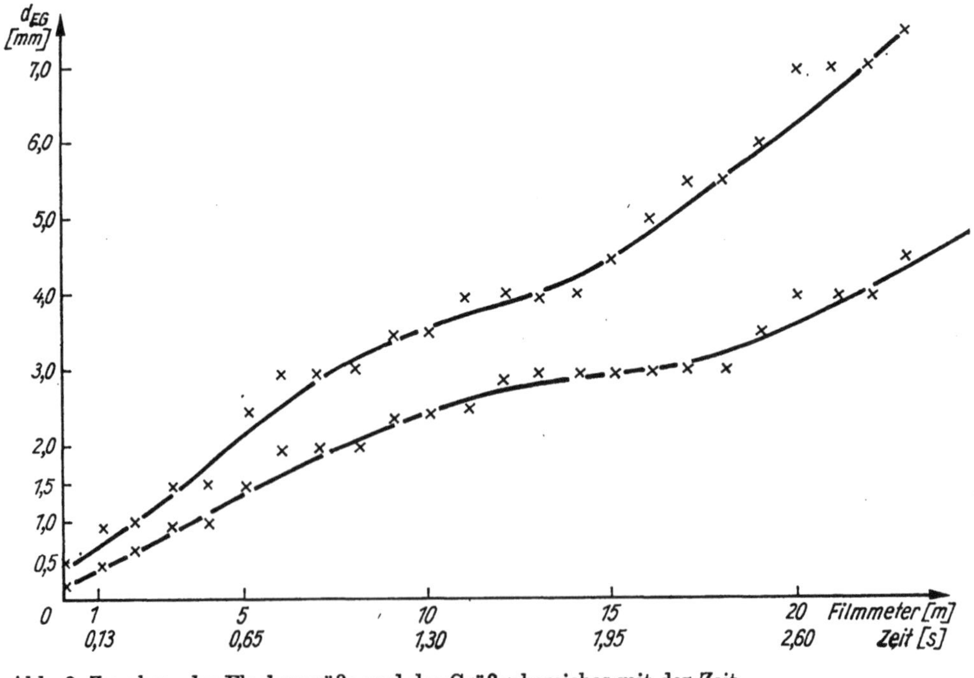

Abb. 3. Zunahme der Flockengröße und des Größenbereiches mit der Zeit

Die verwendete Filmgeschwindigkeit betrug 1000 Bilder s^{-1}. Aus der Darstellung wird eine sehr schnelle Reaktion der Flocken auf veränderte Strömungsverhältnisse sichtbar. Bereits nach ca. 1 Sekunde haben die Flocken das annähernd 6fache ihrer Größe durch Koaleszenz erreicht.

Theoretische Ableitungen

Aus den Untersuchungsergebnissen kann abgeleitet werden:

- Die Flocken reagieren sehr schnell auf Veränderungen der Strömungsgeschwindigkeit und damit der Scherkräfte und des Turbulenzgrades.
- Da in einem Reaktor bei Rührerdrehzahlen von 200—3000 U/min die REYNOLD-zahlen zwischen $14 \cdot 10^3$ und $245 \cdot 10^3$ in Rührblattnähe variieren und die Zirkulationszeiten einige Sekunden betragen können [5], ist mit unterschiedlich großen Flocken sowie ihrer ständigen Veränderung zu rechnen.

Es ist deshalb zu prüfen, inwieweit die Flockenbildung eine Behinderung der Versorgung der Hefezellen im Flockeninneren mit Sauerstoff und Nährstoffen zur Folge hat. Analoge Untersuchungen an Pilzen und anderen Mikroorganismen haben diese Annahme bestätigt [6, 7, 8, 9].
Modelle zur Beschreibung solcher Vorgänge können aus der heterogenen Katalyse abgeleitet werden [9, 10, 11]. Dabei wird die Flocke wie ein Katalysatorkorn aufgefaßt, bei dem die aktiven Zentren durch Transport- und Diffusionsprozesse im Korninneren erreicht werden müssen.
Es ist deshalb folgendes Modell für die Sauerstoffversorgung einer Flocke anzunehmen (Abbildung 4):

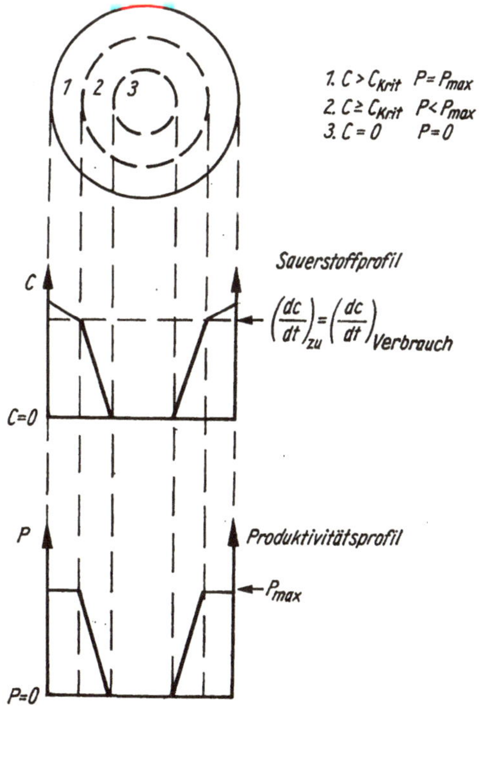

Abb. 4. Zonen des unterschiedlichen Sauerstoffangebotes in einer Flocke.
Korrektur: 2. $C \leq C_{\text{krit}}$ $P < P_{\max}$

Die Flocken werden über einen normalen Diffusionsprozeß mit Sauerstoff versorgt (Turbulenzdiffusion außerhalb der Flocke, Molekulardiffusion im Inneren).
Wenn der Flockendurchmesser einen bestimmten Wert übersteigt, existieren die Hefezellen im Inneren unter Sauerstoffmangel, während in den Randzonen ihre Versorgung gesichert ist.
Dadurch ist ein Sinken der Wachstumsgeschwindigkeit und damit der Produktivität festzustellen, obwohl im ungebundenen wäßrigen Fermentationsmedium eine ausreichende Gelöstsauerstoffkonzentration festzustellen ist.
Um die Geschwindigkeit des Sauerstoffverbrauches und der maximal zulässigen Flokkengröße zu ermitteln, werden aus der spezifischen Sauerstoffverbrauchsgeschwindigkeit die Auszehrzeiten für verschiedene Gelöstsauerstoffkonzentrationen bestimmt.
Die spezifische Sauerstoffverbrauchsgeschwindigkeit ist das Produkt aus spezifischem Sauerstoffverbrauch ($Y_{O_2}^{-1}$) und maximaler spezifischer Wachstumsgeschwindigkeit (μ_{max})

$$\frac{1}{x}\frac{dc}{dt} = Y_{O_2}^{-1} \cdot \mu_{max} = 750 \text{ mg } O_2/\text{g HTS} \cdot \text{h}$$

x = Zellkonzentration

$\frac{dc}{dt}$ = Sauerstoffverbrauchsgeschwindigkeit

$Y_{O_2}^{-1} = 2{,}3$ g O_2/g HTS

$\mu_{max} = 0{,}33\ h^{-1}$

Bei einer Produktivität von ca. 4 g HTS/kg ·h und einer Zellkonzentration von 20 g/kg können die in der Tabelle 1 angegebenen Zehrzeiten bestimmt werden.

Tabelle 1. Sauerstoffzehrungszeiten bei verschiedenen Gelöstsauerstoffkonzentrationen

Sauerstoffgehalt		Zehrungszeit
[mg O_2/l]	[% der Sättigung]	[s]
7,50	100	1,80
6,75	90	1,62
6,00	80	1,44
3,75	50	0,89
2,25	30	0,54
1,50	20	0,36
0,75	10	0,18

Legende: Es wird ein flockenfreies System angenommen. Die Konzentration 20 g HTS soll einer Konzentration von 20 g/l wäßriger Phase entsprechen.

Aus der Tabelle werden sehr große Geschwindigkeiten sichtbar, die sich in sehr kleinen Zeiten für den Verbrauch des angebotenen Sauerstoffes äußern.
Um Aussagen über die annähernd möglichen Verhältnisse in einer Flocke zu erhalten, sind ihre Zusammensetzung und die sich daraus ableitenden Sauerstoffzehrzeiten zu ermitteln. Eine exakte Analyse der Flockenzusammensetzung ist nicht möglich. Deshalb soll die Zusammensetzung der aufschwimmenden Hefe/Öl/Luft/Wassersuspension verwendet werden. Diese Suspension besteht aus 3% Hefe, 72% Wasser und 25% Kohlenwasserstoffe, wenn der Luftinhalt nicht berücksichtigt wird.
Bei einer mittleren Flockengröße von kugelförmiger Gestalt und einem Durchmesser von 3 mm beträgt das Volumen 14,12 mm³ und die Hefekonzentration 0,424 mg.

Der maximal mögliche Sauerstoffgehalt der wäßrigen Phase beträgt $73{,}6 \cdot 10^{-6}$ mg. Dieser Wert ist deshalb ein Maximalwert, da er bei einer Sättigungskonzentration von 100% bestimmt wurde und durch die Vernachlässigung der Gasphase die gebundene wäßrige Phase immer kleiner als angenommen ist.
Diese Sauerstoffmenge wird in der Flocke in einer Zeit von 0,86 s verbraucht, wenn keine Nachlieferung erfolgt.
Da im Inneren der Flocken andere Diffusionsgeschwindigkeiten als im umgebenden Fermentationsmedium anzunehmen sind, soll der Anteil der Hefezellen im Flockeninneren sowie derjenige, der an der Oberfläche existieren kann, bestimmt werden. Dazu wird eine Gestalt der Hefezellen angenommen, die der eines Quaders ähnlich und durch die Ausmaße $3 \times 5\ \mu$ (10^{-6} m) festgelegt ist.
Es wird eine sehr kleine Flocke von 0,3 mm Durchmesser angenommen. Die Dichte soll 0,5 g/cm³ betragen; 50% des Volumens wird von der Gasphase eingenommen. Die Zusammensetzung der wäßrigen und festen Phase entspricht den bereits angegebenen Werten.
Eine solche Flocke enthält ca. $50 \cdot 10^2$ Hefezellen. Die Dichte der Hefe wird mit 1 angenommen. Auf der Oberfläche dieser Flocke können ca. 3×10^2 Hefezellen erwartet werden, wenn die Hefezellen mit ihrer größten Oberfläche auf der Flocke angeordnet sind.
Da eine regellose und zum anderen keine vollständige Belegung angenommen werden muß, sind die ermittelten Zahlen als Durchschnittswerte anzusehen.
Diese Hefezellenmenge stellt 5,0—10% der vorhandenen Zellmenge dar. Daraus wird sichtbar, daß ca. 90% der Hefen sich im Flockeninneren aufhalten.
Welche Unterschiede zwischen den Flockeninneren und den Verhältnissen des umgebenden Fermentationsmediums möglich sind, wird aus dem Vergleich der Diffusionsgeschwindigkeiten sichtbar.
Aus der spezifischen Sauerstoffverbrauchsgeschwindigkeit für Hefen und einer angenommenen aktuellen Sauerstoffkonzentration von 10% des Sättigungswertes ergibt sich eine Zehrgeschwindigkeit von 0,18 Sekunden.
Wenn eine Flockengröße von 0,3 mm angenommen wird, dann muß die Zufuhrgeschwindigkeit des Sauerstoffes zum Flockeninneren

$$r_{O_2 Zu} \geqq 0{,}15\ \text{mm}/0{,}18\ \text{s} \qquad \text{oder}$$

$$r_{O_2 Zu} \geqq 3\ \text{m/h}$$

betragen.

Die angenommene Flocke besitzt eine Oberfläche von 0,283 mm² und eine Schichtdicke entsprechend dem Radius von 0,15 mm.
Der Diffusionskoeffizient für den Übergang des Sauerstoffes aus der Gasphase in die flüssige Phase beträgt $D = 3 \cdot 10^{-5}$ cm²/s oder $10{,}8 \cdot 10^{-6}$ m²/h [12].
Wenn eine Flüssigkeitsschicht von 0,15 mm durchdrungen werden soll, beträgt die Stoffübergangsgeschwindigkeit $\beta = 0{,}7 \cdot 10^{-1}$ m/h.
Aus dem Vergleich der Verbrauchs- mit der Zufuhrgeschwindigkeit kann abgeleitet werden, daß eine ausreichende Versorgung der Hefezellen im Flockeninneren nicht zu erwarten ist.
Dadurch ergeben sich für die Gestatung des Fermentationsprozesses folgende Zielstellungen:

— Auflösen der Flocken durch Senken der Oberflächenspannung oder
— Prozeßführung mit sehr kleinen Flocken, die mit hoher Geschwindigkeit durch Turbulenzzonen oder Scherfelder zur schnellen Sauerstoffnachlieferung gefördert werden müssen.

Eingegangen: 25. 2. 1981

Literatur

[1] Glombitza, F.: Untersuchung zum Übergang des Sauerstoffes in die Zelle im System: Dieselkraftstoff/wäßrige Nährlösung/Luft/Hefe; Dissertation zur Promotion A, Karl-Marx-Universität Leipzig, 1969.
[2] Prokop, A. and Erickson, L. E.: Biotech. Bioeng. **XIII** (1971) 241.
[3] Erickson, L. E., Prokop, A. and Ludvik, M.: Biotech. Bioeng. **XIV** (1972) 587.
[4] Cora, J.: Untersuchung zum Einfluß der Grenzflächenspannung auf die Verhefung von Erdölprodukten; Diplomarbeit, Karl-Marx-Universität Leipzig, 1966.
[5] Sterbaček, Z. and Tausk, P.: Mixing in the chem. Industry; Pergamon Press, 1965.
[6] Yano, T. and Kodama, T.: AGR. Biol. Chem. **25** (1961) 580.
[7] Mueller, J. A.: Biotech. Bioeng. **X** (1968) 331.
[8] Mueller, J. A.: Oxygen Diffusion through a pure Culture Floc of Zoogloca Ramigera: Ph. D. Thesis, Univ. Of. Wisconsin (1966).
[9] Furusawa, T. and Smith, J. M.: Aiche Journal **19** (1973) (2) 401.
[10] Atkinson, B. and Daoud, I. S.: Trans-Instn. Chem. Engrs. **48** (1970) 245.
[11] Tsao, G.: Biotech. Bioeng. **XI** (1969) 1071.
[12] Landolt/Börnstein: Zahlenwerte aus Physik, Chemie, Astronomie, Geophysik, Technik; II. Band 5. Teil (Bandteil b) s. 567, Springerverlag Berlin/Göttingen/Heidelberg, 1960, 6. Auflage.

Acta Biotechnologica 2 (1982) 1, 51–58

Об отношении критической концентрации кислорода к кинетике процесса дыхания

И. Г. Минкевич, В. Д. Кувшинников, В. К. Ерошин

Институт биохимии и физиологии микроорганизмов АН СССР, 142292 г. Пущино, Московской области

Резюме

Рассмотрены методики экспериментального получения кривых потребления кислорода. Проанализирован общепринятый способ нахождения критической концентрации кислорода (*ККК*) по этим кривым. Выявлены те условия эксперимента, при которых кривая потребления кислорода правильно отражает скорость дыхания культуры в ферментере. Показано, что если скорость дыхания в измеряемом диапазоне pO_2 определяется одним и тем же ферментом, то величина *ККК* не является точной характеристикой формы зависимости скорости дыхания от концентрации кислорода.

Summary

This paper considers techniques of measurement of the curves of oxygen consumption by microorganisms. The widely applied method of obtaining the value of the critical oxygen concentration (*COC*) using these curves has been analysed. The experimental conditions necessary for the adequate measurement of the culture respiration rate in a fermenter have been found. It has been shown that in the case when the respiration rate within the considered range of pO_2 is determined by one and the same enzyme, the *COC* value is not an apropriate characteristic of the mode of the respiration rate dependence on oxygen concentration.

Введение

Критическая концентрация кислорода (*ККК*) служит параметром, используемым в исследовательской практике и при промышленном выращивании микроорганизмов для оценки кинетики потребления кислорода [1, 2, 3]. Понятие *ККК* употребляется также при исследовании митохондрий [4], растений [5]. Обычно величина *ККК* определяется по записи полярографического измерения содержания кислорода в среде, когда он потребляется соответствующим биологическим объектом. Считается, что скорость потребления O_2 равна максимальной величине в том диапазоне pO_2, где снижение pO_2 во времени происходит прямолинейно. Точка, в которой pO_2 отклоняется от прямой, считается соответствующей критической концентрации кислорода (см. цит. лит., а также рис. 2, 4, 5). Из-за легкости графического способа своего определения величина *ККК* как параметр выглядит более привлекательно, чем, например, константа полунасыщения K_m. В настоящем сообщении приводятся и обсуждаются данные, указывающие на существование серьезных трудностей при измерении и интерпретации параметра ККК.

Материалы и методы

Эксперимент. Дрожжи *Saccharomyces cerevisiae* штамм 14 выращивались в режиме хемостата в ферментере с рабочим объемом 1,3 л на среде 11Е [6]. В качестве органического субстрата использовали этанол в концентрации 1 об. %. Скорость разбавления составляла 0,05 $час^{-1}$, pO_2 поддерживали автоматически на уровне около 90% от насыщения, pH 4,5, температура 29 °С. Измерения pO_2 и скорости дыхания производились с помощью специальной ячейки, снабженной pO_2-электродом и мешалкой, соединенной с ферментером с помощью петли (рис. 1.). Начало и конец петли находились ниже уровня жидкости в ферментере. Прокачивание культуральной жидкости с клетками проводилось высокопроизводительным насосом, так что время прохождения жидкости от ферментера до ячейки составляло 0,2 сек и pO_2 в ячейке не отличалось от pO_2 в ферментере. Конструкция ячейки обеспечивала быстрое удаление пузырьков воздуха при остановке прокачивания, вследствии чего практически мгновенно прекращалась доставка кислорода в культуральную жидкость, находящуюся в ячейке. Постоянная времени датчика pO_2 была равна 5—8 сек для 95% ответа. Поэтому инерционность датчика не вносила заметной ошибки при концентрации биомассы порядка нескольких г/л по сухому весу.

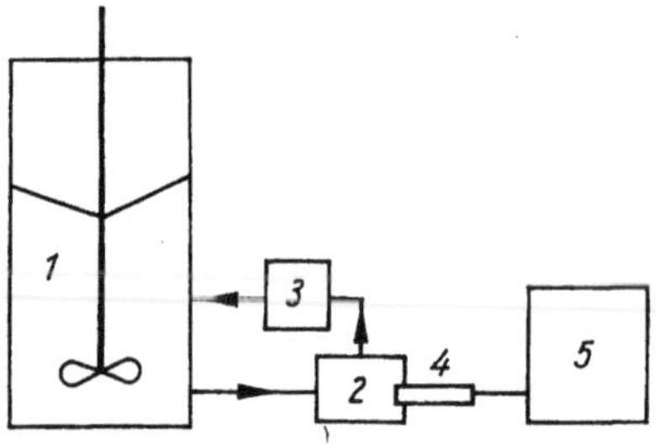

Рис. 1. Схема установки для измерения потребления кислорода микроорганизмами. 1 — ферментер, 2 — ячейка, 3 — насос, 4 — датчик pO_2, 5 — регистрирующий прибор.

При использовании этанола в качестве органического субстрата весовые количества потребленных этанола и кислорода близки друг к другу. Поэтому полное потребление кислорода из среды связано с уменьшением концентрации этанола на величину около 7 мг/л. Остаточная концентрация этанола в культуральной жидкости для данной культуры в режиме хемостата составляла около 40 мг/л. Поэтому изменение скорости потребления кислорода при снижении его концентрации в изолированном образце не связано с изменением концентрации этанола.
Потребление кислорода клетками дрожжей измерялось двумя способами.

1. Выключалась аэрация в ферментере, и одновременно обороты мешалки снижались до 100 об/мин. Ячейка работала в режиме постоянного прокачивания. Этим предотвращалось осаждение клеток на дно ферментера о обеспечивался необходимый проток у мембраны электрода. Пузырьки воздуха, сколько-нибудь заметные наглаз полностью выходили из культуральной жидкости в течение 1—2 сек после отключения аэрации. Регистрировалось падение pO_2 во всем объеме ферментера.
2. Аэрация и перемешивание в ферментере поддерживались на уровне, соответствующем режиму хемостата (pO_2 около 90%). Проток через ячейку периодически останавливался, пузырьки удалялись и регистрировалось падение pO_2 в ячейке с течением времени.

Теория. Рассматривался процесс потребления кислорода, описываемый уравнением Хилла [7]:

$$\frac{dC}{dt} = -V(C) = -\frac{V_m C^n}{K_m^n + C^n}, \tag{1}$$

где C — концентрация кислорода, t — время, $V(C)$ — скорость потребления O_2, K_m — концентрация полунасыщения, V_m — максимальная скорость реакции, n — коэффициент Хилла. Зависимость концентрации кислорода от времени находилась в виде обратной функции $t(C)$ интегрированием уравнения (1):

$$t = \frac{1}{V_m}\left(C_0 - C + K_m \ln \frac{C_0}{C}\right) \quad \text{при } n = 1, \tag{2}$$

$$t = \frac{1}{V_m}\left[C_0 - C + \frac{K_m^n}{n-1}\left(\frac{1}{C^{n-1}} - \frac{1}{C_0^{n-1}}\right)\right] \quad \text{при } n \neq 1, \tag{3}$$

где C_0 — концентрация кислорода в момент времени $t = 0$, когда начинает регистрироваться процесс его потребления. Обычно в эксперименте значение C_0 соответствует $pO_2 = 50-100\%$ от насыщения. При численных расчетах теоретических кривых потребления кислорода по уравнениям (1) и (2)) величина константы полунасыщения полагалась равной $K_m = 0{,}05\ C_0$, то-есть K_m порядка 10^{-5} молей O_2 на литр. Это значение лежит в диапазоне реальных величин K_m, так что рассчитанные кривые вполне могут быть получены в эксперименте.

Результаты и обсуждение

1. *Проблемы измерения.* Кривые потребления кислорода микроорганизмами снимаются в отдельных полярографических ячейках. Если исследуются клетки, выращенные в хемостате, когда остачтоная концентрация лимитирующего субстрата мала, то этот субстрат полностью потребляется за время, затраченное на перенос образца в отдельную ячейку. Поэтому в ячейке клетки находятся я условиях, существенно отличающихся от тех, которые существовали в ферментере. Желательно, однако, получить кривую потребления кислорода в тех же условиях, которые имеют место в самом ферментере.

Один из способов, употребляемых, чтобы получить такую кривую и найти с её помощью критическую концентрацию кислорода, состоит в том, чтобы выключить аэрацию в ферментере, оставив перемешивание [8, 9, 10]. Тогда датчик pO_2, установленный в ферментере, дает кривую его потребления микроорганизмами. Такой подход применяется также и для определения массообменных характеристик ферментера.

Для проверки корректности такого метода измерения мы провели его сравнение с измерением потребления кислорода в специальной ячейке (см. Материалы и методы). Ячейка обеспечивает идентичность условий среды в ней и условий в ферментере. Она гарантирует также отделение пузырьков воздуха от среды, тоесть полное прекращение поступления кислорода в среду из газовой фазы. Поэтому в ячейке снижение pO_2 отражает процесс потребления кислорода клетками.

Как видно из рис. 2, кривая, полученная в ферментере (кривая 1), резко отличается от кривой 2, полученной в ячейке. Во-первых, наклон обеих кривых в их начальной стадии разный. Поэтому скорость дыхания даже при исходном pO_2 измеряется в ферментере неверно. Во-вторых, характер снижения pO_2 в ферментере совсем иной, чем в ячейке. Причина этого заключается в наличии слабого поступления кислорода в культуральную жидкость из воздушной подушки над ней, а также из мелких пузырьков, которые не могут быть отделены от жидкости столь же быстро, как в ячейке. При падении pO_2 в среде возрастает движущая сила массообмена, вследствие чего увеличивается скорость поступления кислорода в среду через поверхность жидкости. Поэтому кривая 1 при снижении pO_2 все больше отличается от кривой 2.

Пользуясь кривой 1 и обычным методом измерения *ККК*, можно получить величину *ККК*, равную 50% от насыщения. Однако, как видно из кривой 2, такая величина критической концентрации кислорода не соответствует действительности, поскольку истинная скорость потребления кислорода данной культурой остается постоянной при pO_2 вплоть до 1%.

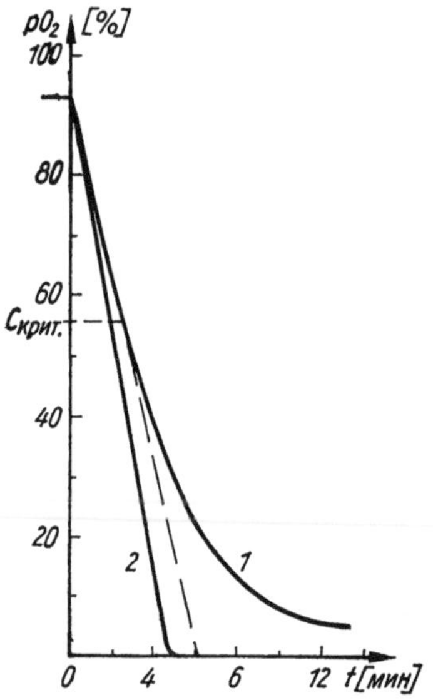

Рис. 2. Кривые снижения содержания кислорода в среде при потреблении его клетками *S. cerevisiae*, полученные способами, указанными в Материалах и методах: 1 — по способу 1; 2 — по способу 2.

Следовательно, измерение надения pO_2 в ферментере с выключенной аэрацией нельзя использовать ни для исследования кинетики потребления кислорода клетками, ни для измерения массообменных характеристик ферментера. В обоих случаях помехой является неполное прекращение аэрации среды, причем интенсивность остаточной аэрации неизвестна. Зарегистрировать падение pO_2, обусловленное только потреблением кислорода клетками, находящимися к тому же в условиях, максимально близких к условиям в ферментере, можно только с помощью специального устройства, обеспечивающего отделение пузырьков воздуха.

2. *Связь ККК с биохимическими механизмами потребления кислорода.* Имея экспериментально полученную кривую зависимости концентрации кислорода в среде от времени, то-есть $C(t)$, можно с помощью несложной математической обработки получить зависимость скорости потребления кислорода клетками $-\frac{dC}{dt} = V(C)$ (в условиях полного отсутствия аэрации, см. выше). Для интерпретации кривой $V(C)$ необходимо качественно охарактеризовать форму этой кривой, используя какие-то численные параметры, и связать ее с возможными биохимическими механизмами потребления кислорода. Рассмотрим, какую поль в связи с этими задачами может играть критическая концентрация кислорода.

Часто имеет место ситуация, когда во всем диапазоне pO_2, где проводились измерения, скорость потребления кислорода определяется осиним и тем же ферментом, действие которого описывается уравнением (1). Тогда форма кривой $V(C)$ полностью описывается тремя параметрами: V_m, K_m, n. С другой стороны, *ККК* также каким-то образом описывает форму кривой $V(C)$. Рассмотрим вопрос о том, дает ли *ККК* дополнительную информацию о $V(C)$.

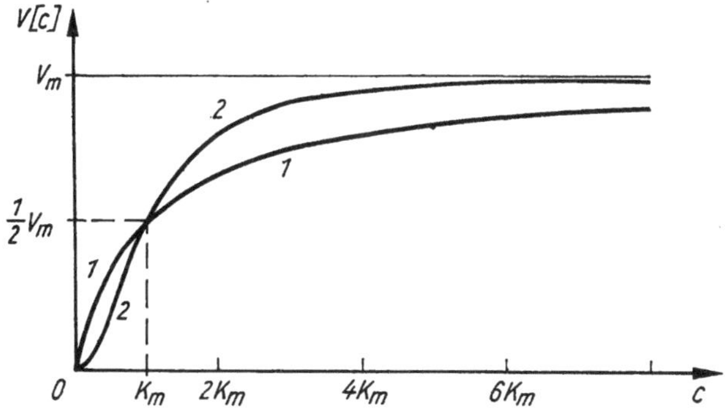

Рис. 3. Зависимость скорости потребления кислорода V от его концентрации c согласно уравнению (1). 1 — $n = 1$ (случай Михаэлиса-Ментен); 2 — $n = 2$.

На рис. 3 представлены кривые зависимости $V(C)$ для $n = 1$ (случай Михаэлиса-Ментен) и $n = 2$ (положительная кооперативность между субъединицами фермента, потребляющего кислород). Обращает на себя внимание, что кривая Михаэлиса имеет затянутый участок приближения к максимальной скорости. При $n = 2$ это приближение происходит гораздо быстрее. Следствия этих различий видны на рис. 4 и 5, где представлены кривые потребления кислорода, рассчитанные по уравнениям

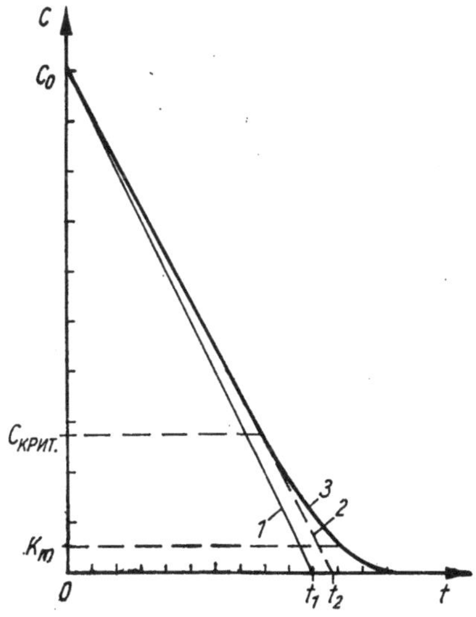

Рис. 4. Кривая потребления кислорода ферментативной системой, действующей по механизму Михаэлиса-Ментен ($n = 1$). Пояснения в тексте.

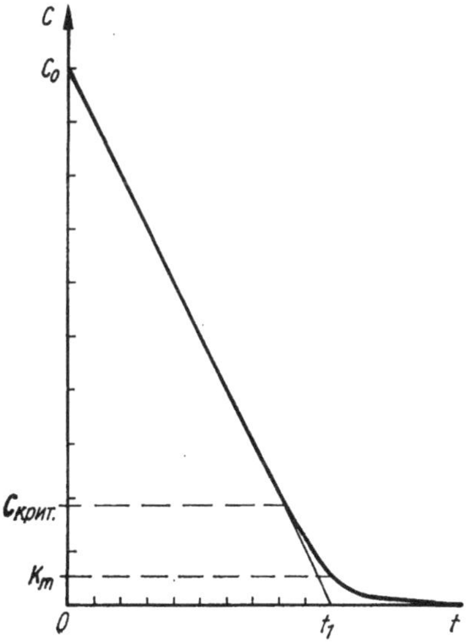

Рис. 5. Кривая потребления кислорода ферментативной системой, у которой коэффициент Хилла $n = 2$. Пояснения в тексте.

соответственно (2) и (3). Именно такие графики должен дать в эксперименте полярографический метод, когда обеспечено отсутствие притока кислорода (см. выше).

Рассмотрим случай кинетики Михаэлиса-Ментен (рис. 4). Прямя 1 изображает снижение содержания кислорода в среде, если бы его потребление проходило все время с максимальной скоростью V_m. Тогда исчерпание кислорода происходило бы в момент $t_1 = C_0/V_m$. Однако, если предположить, что максимальная скорость потребления кислорода соответствует прямолинейному участку $C(t)$ и использовать линейку, как это принято в лабораторной практике, то максимальная скорость будет приписана другой прямой, которая на рис. 4 начинается в точке C_0, а кончается в точке t_2 (прямая 2). Заметное отклонение кривой $C(t)$ от прямой 2 начинается при концентрации кислорода, которая, согласно общепринятой методике, должна рассматриваться как критическая ($C_{\text{крит.}}$ на рис. 4). В действительности же участок кривой $C(t)$ между концентрациями C_0 и $C_{\text{крит.}}$ не прямолинеен, но имеет очень слабую кривизну. Отклонение его от прямой 1 вызвано именно тем, что кривая $V(C)$ при $n = 1$ очень медленно приближается к своему максимальному значению V_m (см. рис. 3). Таким образом, неточности самописца и неточности оценки линии ,,на-глаз" приводят, прежде всего, к ошибочному определению величины V_m. По этим же причинам величина $C_{\text{крит.}}$ также может быть найдена со значительным разбросом. Неточность определения V_m снижается при более высоких значениях n. Для $n = 2$ различие между прямыми 1 и 2, имевшее место на рис. 4, пропадает, и точно приложенная линейка (при условии, что самописец сделал идеальную запись) дает значение $t_1 = C_0/V_m$, откуда легко вычисляется V_m (рис. 5). Это — следствие более резкого приближения кривой $V(C)$ KV_m при $n = 2$ (см. рис. 3). Однако место отхода кривой $C(t)$ от прямой по-прежнему определяется неточно.

Основной же недостаток величины $C_{\text{крит.}}$, на наш взгляд, заключается даже не в возможных ошибках, возникающих при столь субъективном способе нахождения этой величины. Считается, что $C_{\text{крит.}}$ характеризует форму зависимости скорости реакции от концентрации субстрата (в данном случае кислорода), а именно, она является такой концентрацией, что при более высоких концентрациях субстрата скорость его потребления практически постоянна, а при более низких эта скорость заметно падает. Создается впечатление, что при концентрации кислорода, равной $C_{\text{крит}}$., происходит какое-то резкое изменение хода кривой $V(C)$ (напомним, что мы рассматриваем случай, когда во всем изучаемом диапазоне pO_2 узкое звено в дыхательной цепи не меняется). В действительности же при $C = C_{\text{крит.}}$ никаких резких изменений на кривой $V(C)$ не наблюдается. Из рис. 4 и 5 видно, что $C_{\text{крит.}}$ значительно выше, чем K_m (для $n = 1$ и $n = 2$ в 4—5 раз). В этом диапазоне концентраций субстрата кривые $V(C)$ находятся уже вне области сильного спада. Никакого излома этих кривых, резкого перехода от постоянства к спаду не наблюдается. Таким образом, $C_{\text{крит.}}$ гораздо в большей степени характеризует форму кривой $C(t)$, чем $V(C)$.

Из сказанного следует, что критическая концентрация кислорода служит весьма грубой оценкой формы кривой $V(C)$ и не может претендовать на роль параметра. Если параметр K_m участвует в формуле (например, [1]), которая описывает всю кривую $V(C)$, то $C_{\text{крит.}}$ лишь указывает диапазон, внутри которого находится K_m, и не дает сколько-нибудь точной количественной характеристики скорости дыхания. Поэтому $C_{\text{крит.}}$ содержит гораздо меньше информации, чем K_m. Если величина K_m определена, то не имеет смысла пользоваться $C_{\text{крит.}}$ или сравнивать эти величины. Очевидно, это справедливо не только для кинетики Хилла, но и для любого вида ферментативной кинетики, поскольку кривые $V(C)$ для одного фермента обычно не имеют изломов.

Восможны, однако, случаи, когда величина ККК дает информацию о кинетике по требления кислорода, дополнительную к информации, заключенной в K_m. Это имеет место, если при снижении pO_2 происходит переключение узкого звена в дыхательной цепи. Такая форма $V(C)$ показана на рис.6. При $C > C_{\text{крит.}}$ скорость дыхания определяется ферментом с нузким K_m, при $C < C_{\text{крит.}}$ — другим ферментом с высоким K_m. Переключение узкого места происходит при $C = C_{\text{крит.}}$. Тогда кривые $V(C)$ и $C(t)$ характеризуются изломом при $C = C_{\text{крит.}}$ Наличие и место излома можно установить приблизительно, используя линейку и запись $C(t)$. Однако наиболее достоверно убедиться в наличии излома можно только используя статистическую обработку кривой $C(t)$.

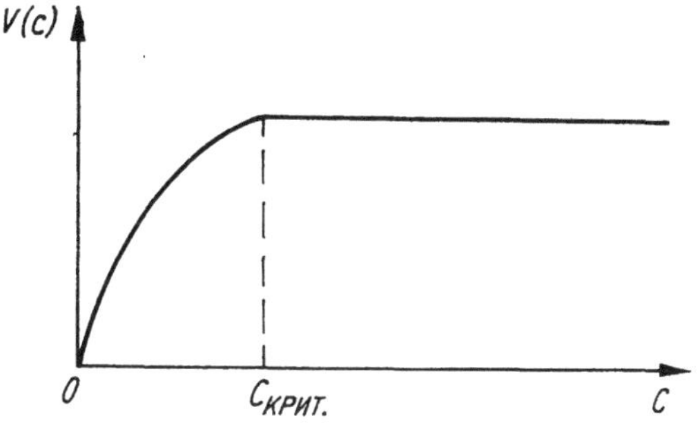

Рис. 6. Зависимость скорости потребления кислорода от его концентрации в в случае, когда при $C = C_{\text{крит.}}$ происходит преключение узкого места в дыхательной цепи.

Выводы

1. При измерении скорости дыхания по кривой потребления кислорода необходимо пользоваться устройством, обеспечивающим быстрый перенос образца из ферментера и полное прекращение аэрации. Ферментер с выключенной барботажной системой не обеспечивает быстрое прекращение аэрации, и поэтому кривая снижения pO_2 в нем не дает достаточно точных данных для измерения скорости дыхания.

2. Влияние концентрации кислорода C на скорость дыхания V непосредственно отражено формой кривой $V(C)$, в то время, как кривая потребления кислорода $C(t)$ дает лишь косвенную информацию об этом влиянии. При получении кривой $V(C)$ из $C(t)$ следует использовать статистическую обработку. Если кривая $V(C)$ имеет гладкую форму (без изломов), то использование KKK дает плохое описание формы $V(C)$, и предпочтительным является использование константы полунасыщения K_m. Если кривая $V(C)$ имеет излом, по одну сторону которого наблюдается резкий спад скорости дыхания, то, напротив, можно рекомендовать использовать величину KKK, рассматривая ее как концентрацию кислорода, при которой происходит излом кривой $V(C)$.

Авторы выражают глубокую благодарность коллективу отдела биоэнергетики ИБФМ АН СССР за плодотворную дискуссию.

Eingegangen: 29. 1. 1981

Литература

[1] Harrison, D. E. F.: J. Appl. Chem. Biotechnol., **22** (1972) 417.
[2] Harrison, D. E. F.: Adv. Microb. Physiol. **14** (1976) 243.
[3] Phillips, D. M., Johnson, M. J.: J. Biochem. Microbiol. Technol. Engineer. **3** (1961) 277.
[4] Stolp, W., Thiemann, V., Weber, D., Weiss, Ch.: Pflügers Archiv, **323** (1971) 250.
[5] Armstrong, W., Gaynard, T. J.: Physiol. Plant. **37** (1976) 200.
[6] Кувшинников, В. Д., Воробьев, А. В., Ерошин, В. К., Минкевич, И. Г.: Прикладн. биохим. микробиол. **14** (1978) 366.
[7] Корниш-Боуден, Э.: Основы ферментативной кинетики, ,,Мир“, М., 1979, стр. 168.
[8] Перт, С. Дж.: Основы культивирования микроорганизмов и клеток, ,,Мир“, М., 1978 стр. 115.
[9] Páca, J., Grégr, V.: J. Appl. Chem. Biotechnol. **27** (1977) 155.
[10] Sonnleitner, B., Heinzle, E., Braunegg, G., Lafferty, R. M.: Europ. J. Appl. Microbiol. Biotechnol. **7** (1979) 1.

Acta Biotechnologica 2 (1982) 1, 59–71

Some Aspects of Fermentation as an Unit Process in Chemical Engineering

F. G. MUELLER

University of Basel / CHEMAP AG,
8712 Männedorf, Switzerland

Summary

There are described possibilities of scale up in stirred tank fermenters being based on known relations of mass-, energy- and heat transfer. The represented mathematical description shows the problems of the necessary scale up in the practice with the theory of similarity in the stirred tank fermenter. Relating on the method of RUSHTON for calculation of the energy capacity in different stirred fermenters the influence of the aeration is examined and confirmed with experimental results of the author. By the system of equations the known parameters of the aerobic fermenters can be calculated.

Zusammenfassung

Dargestellt werden Möglichkeiten der Maßstabsübertragung für den Rührfermentor, aufbauend auf den bekannten Beziehungen für den Stoff-, Energie- und Wärmeübergang. Die abgehandelte mathematische Beschreibung zeigt die Probleme der in der Praxis notwendigen Maßstabsübertragung anhand der Ähnlichkeitstheorie für den Rührfermentor. Ausgehend von der Verfahrensweise von RUSHTON zur Berechnung der Energieaufnahme für verschiedene Rührkessel wird der Einfluß der Belüftung untersucht, mit eigenen Ergebnissen bestätigt und auf die Bestimmung des k_La-Wertes eingegangen. Mit dem gegebenen Gleichungssystem können die bekannten Parameter des aeroben Fermentors berechnet werden.

Unit operations in chemical engineering normally consist of one or several transfers of heat, mass or quantity of motion.

So extraction by solvent of a substance, dissolved in a liquid for instance, is a mass transfer of the solute between the liquids. In fermentation several different masstransfers as well as heat transfer and transfer of motion are involved.

It is the aim of any systematic treatment and mathematical description of unit operations to derive mathematical models, which describe the unit operation with an accuracy allowing scale up from a pilot size to the industrial size. The tool to achieve this is the theory of similitude. As always in mathematics similitude is a matter of definition. This means that we have to define what criteria have to be fullfilled when we say, that the model and full scale plant are similar.

The simplest laws of similitude we all know from geometry. In Fig. 1 we see two triangels, which are similar. The conditions of similitude being:

$$\frac{a_1}{a_2} = \frac{b_1}{b_2} = \frac{c_1}{c_2} = \text{const.}$$

thus a linear similitude.

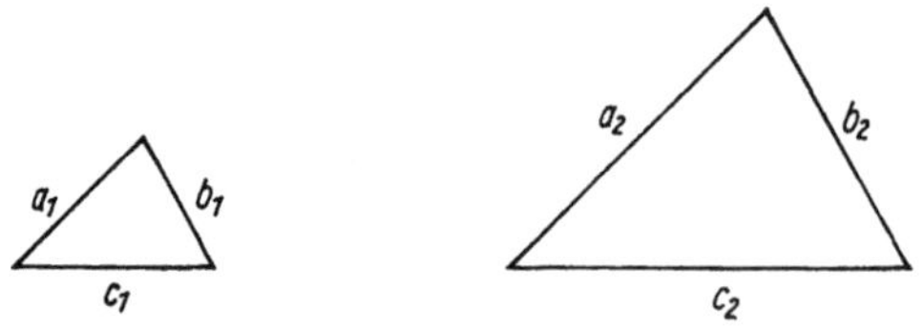

Fig. 1. Geometrical similitude

For a more complicated threedimensional model as in Fig. 2 the conditions of similitude are

$$\frac{X_{P1}}{X_{P2}} = \frac{Y_{P1}}{Y_{P2}} = \frac{Z_{P1}}{Z_{P2}} = \text{const.} \tag{1}$$

and if we allow for shift of the coordinates

$$\frac{X_{P1} - X_{01}}{X_{P2} - X_{02}} = \frac{Y_{P1} - Y_{01}}{Y_{P2} - Y_{02}} = \frac{Z_{P1} - Z_{01}}{Z_{P2} - Z_{02}} = \text{const.} \tag{2}$$

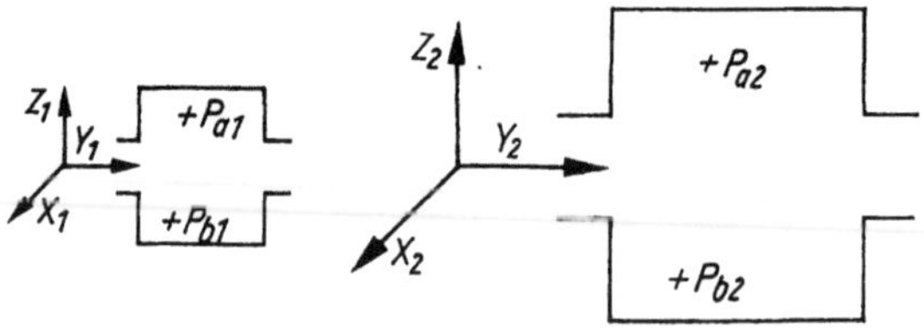

Fig. 2. Three dimensional geometrical similitude

For each point of our model equation 2 has to be fullfilled. Since reactors generally are run through by fluxes we have to define some more similitudes. The input values are for instance U_{02} the liquid velocity, C_{02} the concentration and T_{02} the temperature and the corresponding values in the model are U_{01}, C_{01} and T_{01}.
We define thermal similitude by the equation.

$$\frac{T_{11}}{T_{12}} = \frac{T_{21}}{T_{22}} = \frac{T_{31}}{T_{32}} = \cdots = \frac{T_{n1}}{T_{n2}} = \text{const.} \tag{3}$$

or more generally

$$\frac{T_{11} - T_{01}}{T_{12} - T_{02}} = \frac{T_{21} - T_{01}}{T_{22} - T_{02}} = \cdots = \frac{T_{n1} - T_{01}}{T_{n2} - T_{02}} = \text{const.} \tag{4}$$

Chemical similitude is defined in an analog way.

$$\frac{C_{11}}{C_{12}} = \frac{C_{21}}{C_{22}} = \frac{C_{31}}{C_{32}} = \cdots = \frac{C_{n1}}{C_{n2}} = \text{const.} \tag{5}$$

or again

$$\frac{C_{11} - C_{01}}{C_{12} - C_{02}} = \frac{C_{21} - C_{01}}{C_{22} - C_{02}} = \cdots = \frac{C_{n1} - C_{01}}{C_{n2} - C_{02}} = \text{const.} \tag{6}$$

Hydrodynamical similitude may be defined as follows

$$\frac{U_{11}}{U_{12}} = \frac{U_{21}}{U_{22}} = \frac{U_{31}}{U_{32}} = \cdots = \frac{U_{n1}}{U_{n2}} = \text{const.} \tag{7}$$

Table 1.

Name	Abrevation	Definition	Meaning	Application
ARCHIMEDES	Ar	$\frac{gl^3\varrho(\varrho - \varrho_0)}{\eta^2}$	Gravity/inner friction	Movement of fluids with density differences
BODENSTEIN	Bo	$\frac{uL}{Da}$	Mass in macroscopic motion/ back mixed mass	Mixing in flowing media
EULER	Eu	$\frac{\Delta pk}{\varrho u^2}$	Pressure/inertial force	Friction of fluids in tubes
FRAUDÉ	Fr	$\frac{u^2}{gL}$	inertial force gravity	mixed vessels
GALILEI	Ga	$\frac{L^3 g\varrho^2}{\eta^2}$	gravity/inner friction	Movement of fluids in a field of gravity
LEWIS	Le	$\frac{\lambda}{\varrho CpD}$	Heat conductivity/ Diffusion	simultaneous heat and mass transfer
NUSSELT	Nu	$\frac{\alpha L}{\lambda}$	total heat transfer/ heat conduction	Heat transfer
PRANTL	Pr	$\frac{Cp\eta}{\lambda}$	molecular impulse/ heat transfer	Heat transfer
REYNOLDS	Re	$\frac{Lu\varrho}{\eta}$	impulse/inner friction	Fluids in motion
SCHMIDT	Sc	$\frac{\eta}{\varrho D}$	molecular impulse/ mass transfer	Mass transfer
SHERWOOD	Sh	$\frac{\beta L}{D}$	total mass transfer/ Diffusion	Mass transfer
WEBER	We	$\frac{u^2\varrho L}{\sigma}$	inertial force/ interfacial forces	Bubbles drops

Symboles

C_p	kJ kg^{-1} K^{-1}	specific heat
D_a	m^2 s^{-1}	diffusion coefficient
g	m s^{-2}	acceleration due to gravity
L	m	characteristic length
ΔpK	Ug s^{-2} m^{-1}	pressure loss due to friction
U	m s^{-1}	velocity
α	kg s^{-3} grd	heat transport number
β	m s^{-1}	mass transport number
λ	m kg s^{-3} K	heat conduction number
ϱ	kg m^{-3}	density
ϱ_0	k m^{-3}	density of lighter medium
σ	kg s^{-2}	interfacial tension
η	kg m^{-1} sec^{-1}	viscosity

That is so say the ratios of velocities at geometrically similar points are similar. Here we have to take into account that only similar flows may be compared i.e. similitude is only given between two laminar or between two turbulent flows. A laminar and a turbulent flow never can be similar.
Theory of similitude is based on dimensionless numbers some of them shown in table 1.

The important theorem of the theory of similitude is:

Model and full scale execution show a similar behaviour if under the same limiting values the characteristic dimensionless numbers for model and full scale execution have the same numerical values.
Let us show now how dimensionless numbers are used:

As an example we will derive the equations of similitude of impellers in a stirred vessel such as a fermenter.

For this we will follow Rushton's concept of the power number. A representative velocity u of liquid in an agitated vessel will be proportional to the tip velocity of the impeller.

$$u \approx nD_i \tag{8}$$

with

u = velocity
n = rotation speed of impeller
D_i = impeller diameter

The external force imposed on an unit volume of liquid at a given tip speed will then be

$$\frac{P}{nD_i} \cdot \frac{1}{D_i^3} \tag{9}$$

where

P = power required for agitation

On the other hand the inertial force of an unit volume of liquid in the vessel will be:

$$\frac{\varrho n D_i}{\frac{1}{n}} = \varrho n^2 D_i \tag{10}$$

The viscous force in the liquid will be:

$$\eta \frac{D_i^2}{D_i^3} \cdot \frac{nD_i}{D_i} = \eta \tag{11}$$

provided:

η = liquid viscosity

We now define the dimensionless power number Np as the ratio of external to inertial forces of the liquid.

$$\frac{\text{External force}}{\text{Inertial force}} = \frac{\frac{P}{nD_i4}}{\varrho n^2 D_i} = \frac{P}{n^3 D_i^5 \varrho} = N_P \tag{12}$$

Motions of a liquid in an agitated vessel are depicted by the ratio of inertial to viscous forces of the liquid i.e. a modified REYNOLDS number of the impellers.

$$\frac{\text{Inertial force}}{\text{Viscous force}} = \frac{\varrho n^2 D_i}{\frac{\eta n}{D_i}} = \frac{n D_i^2 \varrho}{\eta} = N'_{\text{Re}} \quad . \tag{13}$$

The power characteristics of liquid agitation with various types of impellers established by RUSHTON are shown in Fig. 3.

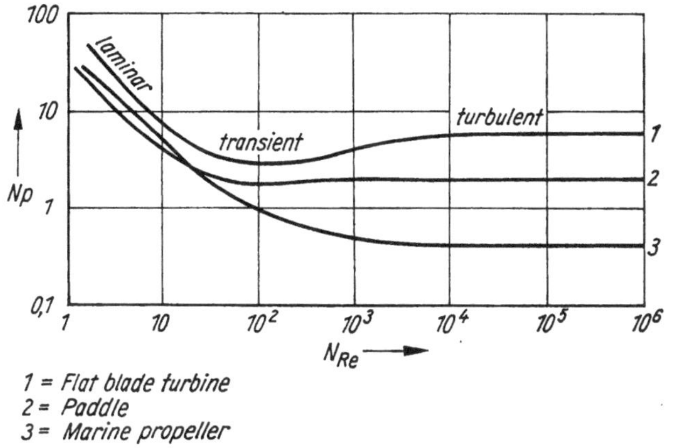

Fig. 3. Power number versus REYNOLDS number RUSHTON: "Power characteristics of impellers"

Once such a diagram is established for a particular type of stirrer it is very simple to calculate power uptake for different sizes of stirred vessels by the following equation:

$$P = \varrho n^3 D_i^5 N_P \quad [\text{W}] \tag{14}$$

If more than one agitator is installed in the same vessel we have to distinguish between baffled and unbaffled vessels. In fermentation normally fully baffled vessels are used. The conditions for fully baffling are given by:

$$\left(\frac{Bw}{D_t}\right)^{\frac{6}{5}} n_B = 0.35 \tag{15}$$

with:

B_W = baffle width [m]
D_T = tank diameter [m]
n_B = number of baffles

Under fully baffled conditions the power uptake of several identical turbines on the same shaft is practically the sum of the power uptake of every single turbine.
This fact is shown by table 2, which gives the results of power measurements in a 1500 l pilot fermenter.
Of course power uptake varies with the aeration rate. To cope with this again dimensionless numbers will help. OHYAMA defined the following function.

$$\frac{P_g}{P} = f(N_a) \tag{16}$$

with

P_g = power uptake under gassed conditions
p = power uptake in pure liquid
N_a = aeration number

The aeration number was defined by OHYAMA et al. as

$$N_a = \frac{\text{air velocity through sectional vessel area}}{\text{tip velocity of the impeller}}$$

thus:

$$\frac{\frac{Q}{D_i^2}}{nD_i} = \frac{Q}{nD_i^3} = N_a \qquad (17)$$

Table 2.

RPM	1 Turbine	2 Turbines	3 Turbines
328	1.24 KW	2.47 KW	3.73 KW
360	1.51 KW	3.1 KW	4.60 KW
388	2.08 KW	3.87 KW	6.07 KW
415	2.47 KW	4.73 KW	7.92 KW

Fig. 4 shows the resulting functions for two types of impellers.
Our own experiments confirmed the practicability of this method, however it has to be stressed that the type of air sparger used, influences the function f considerably.
Let us speak now of oxygen transfer, which is of utmost importance for high productivity aerobic fermentation. We will show here that oxygen transfer is closely related to power input in the liquid, a fact which is often ignored.

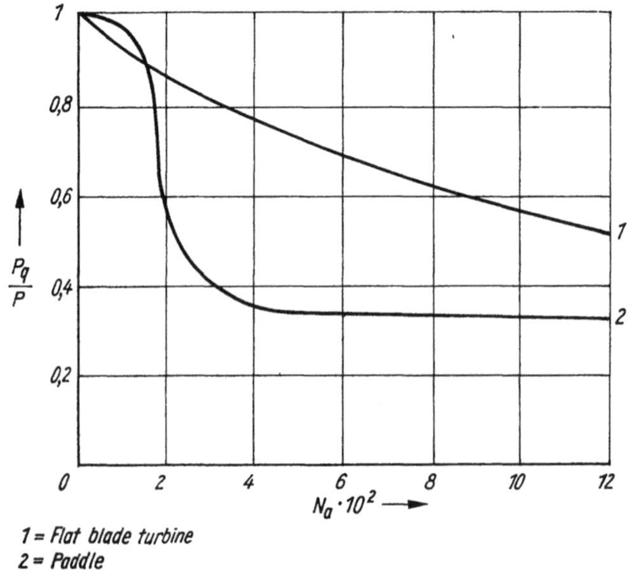

Fig. 4. Power requirements under gassed conditions

Gas liquid transfer is a diffusion process, which in its simplest form can be described by FICK's first law.

$$Z = -D \cdot A \frac{dcv}{dx} \cdot t \tag{18}$$

with:

Z = number of Mols which diffuse into the liquid [dl]
D = diffusion coefficient [m²/sec]
A = aerea through which the diffusion takes place (gas liquid interface) [m²]
$\frac{dcv}{dx}$ = the concentration gradient [Mol/m⁴]
t = time [sec]

What can be done, to get a large diffusion rate? Since D for a given system is a constant and since also dx the diffusion path is beyond our influence, we have but two possibilities. Namely, to make the gas liquid interface and dcv as large as possible. Of course gas transfer in a fermenter from the gas into the microbial cell is much more complex as described by FICK's law, which by the way only describes steady state conditions. The gas has to diffuse first from the gas phase into the liquid phase passing a film at the interface between the gas bubble and the liquid, then it has to diffuse into the microorganisms by a still more complicated process. Fortunately the second diffusion process is normally much faster than the gas liquid transfer so that we can concentrate on the first step.
A practical way to describe oxygen transfer rate (OTR) is the following equation:

$$OTR = k_L a (C_g - C_L) \tag{19}$$

with

OTR = oxygen transfer rate (eg [Mol m^{-3} sec^{-1}])
k_L = mass transfer coefficient for the liquid phase [m sec^{-1}]
C_g = oxygen concentration in the gas phase [Mol m^{-3}]
C_L = oxygen concentration in the liquid phase [Mol m^{-3}]
a = specific interface (m^2/m^3) [m^{-1}]

C_L and C_g may be determined by relatively simple measurements and the OTR may be determined either by chemical methods (sulphite oxidation) or better with a culture of growing microorganisms. With this the important characteristic figure of a fermenter $k_L a$ may be determined.
However there are also quite reliable methods of estimating k_L by calculation. We give here without derivation, the equation of CALDERBANK.
He finds for the gas liquid masstransfer coefficient k_L $(g - L)$:

$$k_L = 0.31 \left(\frac{\eta_L}{\varrho_L D_L}\right)^{-\frac{2}{3}} \cdot \left(\frac{\eta_L g}{\varrho_L}\right)^{\frac{1}{3}} \tag{20}$$

with:

η_L = viscosity of liquid (kg m^{-1} sec^{-1})
ϱ_L = liquid density (kg m^{-3})
D_L = coefficient of diffusion (m^2 sec^{-1})
g = acceleration due to gravity (9.81 m sec^{-2})

The next parameter to be calculated is a. This becomes possible by the following approach.

Assuming the gas bubbles to be of spherical shape the volume of a gas bubble is

$$V_b = \frac{\pi}{6} d^3$$

and the surface

$$A_b = \pi d^2$$

since a is a product of all bubble surfaces A_b times the number N of bubbles retained in the total liquid of the stirred tank divided by the ungassed volume, and V_b times the number N of bubbles divided by the ungassed volume equals the gas hold up H (difference in volume between gassed and ungassed state) divided by the liquid volume in the ungassed state (dimensionless) the following equation holds true:

$$\frac{H}{a} = \frac{d}{6} \qquad a = \frac{6H}{d} \text{ [m}^{-1}\text{]} \tag{21}$$

This leeds to a very elegant method to determine a provided that d (mean SAUTER diameter) is known. It may either be measured by different methods or calculated by using one of the equations given below.
The simplest equation, which according to our experience gives quite correct values, is the equation of LEE and MEYRICK:

$$d = 3.15 \frac{\gamma^{\frac{3}{5}}}{\left(\frac{P}{V}\right)^{\frac{2}{5}} \varrho_L^{\frac{1}{5}}} \text{ [m]} \tag{22}$$

Some times a still better fit is achieved with the equation of CALDERBANK.

$$d = 2.25 \left(\frac{\gamma^{\frac{3}{5}}}{\left(\frac{P}{V}\right)^{\frac{2}{5}} \varrho_L^{\frac{1}{5}}} \right) H^{\frac{2}{5}} \left(\frac{\eta_g}{\eta_L}\right)^{\frac{1}{4}} \text{ [m]} \tag{23}$$

With the given set of equations the most important parameter of an aerobic fermenter may be calculated and with this it seems that scale up — at least of a stirred tank fermenter — should not be to much of a problem.
But unfortunately as mostly in engineering, this is only one part of the truth.
A simple example will show this. For an unagitated aerated vessel like for instance a bubble column the following simplified equation is given by HUMPHREY:

$$k_L a = \text{const} \frac{Q}{V} H_L^{\frac{2}{3}} \frac{1}{d} \text{ [sec}^{-1}\text{]} \tag{24}$$

with:

Q = aeration rate [m³/sec]
V = ungassed liquid volume [m³]
H_L = height of liquid [m]

If we choose now a scale up concept of constant $k_L a$ we may write

$$\frac{(k_L a)_1}{(k_L a)_2} = \frac{\left(\frac{Q}{V}\right)_1}{\left(\frac{Q}{V}\right)_2} \cdot \left(\frac{H_{L1}}{H_{L2}}\right)^{\frac{2}{3}} \cdot \frac{d_1}{d_2} \tag{25}$$

Since the left hand term of equation (25) equals unity and if we assume that the bubble diameter is not affected by the liquid height we may write

$$\frac{\left(\frac{Q}{V}\right)_2}{\left(\frac{Q}{V}\right)_1} = \left(\frac{H_{L2}}{H_{L1}}\right)^{-\frac{2}{3}} \tag{26}$$

From this follows that, if we scale up a geometrical similar fermenter by a volumetric factor of 100, thus by a linear factor of 4.64, and if we needed in our small fermenter a an aeration rate of 1 vvm, we would need in our large fermenter $4.64^{-2/3} = 0.36$ vvm. This seems at a first glance logical, since the bubble residence time becomes longer and thus the oxygen yield better and very encouraging since the smaller aeration rate will make the fermentation more economical.
But let us assume now, that we used four our pilot experiment a fermenter, which at 1 vvm due to its very efficient EFFIGAS®-stirrer, gives an oxygen yield of about 65%. This would result in an oxygen yield of $\frac{65}{0.36} = 180\%$, which is of course nonsense.
The example — however calculated with a simplified equation, which strictly does not hold true for an agitated vessel — shows that scale up of a fermenter is quite complex. Since according to the equations of LEE and MEYRICK and CALDERBANK, the power input per unit of volume has a great influence on the bubble diameter and since the latter is very important for the gas liquid exchange interface a scale up concept of constant power input per unit of volume is very promising.
Dimensional analysis of the mass transfer coefficient in agitated and geometrically similar systems leeds to the following expression:

$$\frac{k_L D_i}{D} \approx \left(\frac{D_i^2 n \varrho}{\eta}\right)^{\alpha} \cdot \left(\frac{\eta}{\varrho D}\right)^{\gamma} \tag{27}$$

where:

k_L = mass transfer coefficient
D_i = impeller diameter
D = molecular diffusivity (of the substance under consideration)
n = rotation speed of impeller
η = liquid viscosity
ϱ = liquid density
$\alpha; \gamma$ = empirical exponents

Equation (27) can also be used for calculating heat transfer under agitated conditions if the SHERWOOD number

$$\text{Sh} = \frac{k_L D_i}{D}$$

is replaced by the NUSSELT number

$$\text{Nu} = \frac{h D_i}{k}$$

where:

h = film coefficient of heat transfer
k = thermal conductivity

and the SCHMIDT number

$$\mathrm{Sc} = \frac{\eta}{\varrho D}$$

by the PRANTL number

$$\mathrm{Pr} = \frac{c_p \eta}{k}$$

with C_p = specific heat
As to achieve the same numerical value of k_L in two geometrically similar vessels we derive from (27)

$$\frac{n_1}{n_2} = \left(\frac{D_{i2}}{D_{i1}}\right)^{\frac{2\alpha-1}{\alpha}} \tag{28}$$

where the index 1 means the small scale and index 2 the large scale equipment.
For the calculation of the power uptake of a stirrer in a baffled vessel we derived:

$$p \approx n^3 D_i^5$$

from this we get

$$\frac{P_2}{P_1} = \left(\frac{n_2}{n_1}\right)^3 \cdot \left(\frac{D_{i2}}{D_{i1}}\right)^5 \tag{29}$$

by combining (28) and (29)

$$\frac{P_2}{P_1} = \left(\frac{D_{i2}}{D_{i1}}\right)^{5-3\left(\frac{2\alpha-1}{\alpha}\right)} \tag{30}$$

and finally

$$\frac{\left(\frac{P_2}{V_2}\right)}{\left(\frac{P_1}{V_1}\right)} = \left(\frac{D_{i2}}{D_{i1}}\right)^{2-3\left(\frac{2\alpha-1}{\alpha}\right)} \tag{31}$$

Since for many fermentations the concept of equal specific power input works quite good, one may assume that the coefficient for mass transfer of oxygen in fermenters is around 0.75, which value makes the right hand side of equation (31) equal unity.
If this is true the concept is applicable and we can write

$$\frac{P_1}{N_1{}^3 D_i{}^5} = \frac{P_2}{N_1{}^3 D_i{}^5} \quad D_i \approx D_t$$

or since D_i is proportional to D_t we substitute D_i by D_t and multiply numerator and denominator of the respective fractions with the respective D_t thus getting

$$\frac{P_1}{D_{t1}^3} \frac{D_{t1}}{N_1{}^3 D_{t1}^3} = \frac{P_2}{D_{t2}^3} \frac{D_{t2}}{N_2{}^3 D_{t2}^3}$$

because

$$\frac{P_1}{V_1} = \frac{P_2}{V_2} = \text{const.}; \qquad V \approx D_t{}^3$$

is constant we get

$$\frac{1}{N_1{}^3 D_{t1}^2} = \frac{1}{N_2{}^3 D_{t2}^2}$$

or

$$\frac{N_1{}^3 D_{t1}^3}{N_2{}^3 D_{t2}^3} = \frac{D_{t1}}{D_{t2}}$$

from this follows

$$\frac{N_1 D_{t1}}{N_2 D_{t2}} = \left(\frac{D_{t1}}{D_{t2}}\right)^{\frac{1}{3}} \tag{32}$$

Since

$$\frac{N_1 D_{t1}}{N_2 D_{t2}}$$

is the ratio of the tip speeds of the impellers this means that the tip speed, which is a measure for the liquid velocity in an agitated vessel, increases with the third root of the ratio of the vessel diameters. As an example, if we take our volumetric scale up of 100, which means a linear scale up of 4.64, we get

$$N_2 D_{i2} = N_1 D_{i1}\,(4.64)^{\frac{1}{3}}$$

or with other words our tip speed has to increase by a factor of 1.67.
It has to be stressed here that the increase of the tip speed also increases the shear forces to which the organism is subjected. It is thus evident that this concept cannot always be used if very large scale up ratios are necessary. Especially filamentous fungi may be very sensitive to an increase in shear.
An other very important scale up factor is the mixing time t. It is defined as the period of time necessary for a liquid droplet brought into an agitated vessel filled with a liquid of the same physical properties as the droplet to be uniformely mixed with the bulk liquid. Mixing times may be measured by very simple methods, for instance by adding acid, base, electrolytes or dyestuffs and measuring uniformity by a suitable detector at various points in the reactor. Mixing characteristics of a stirrer system may be depicted by a similar approach as power uptake (Rushton-method) Norwood et al. defined a mixing time factor ϕ as

$$\phi = \frac{t(nD_i^2)^{\frac{2}{3}} \cdot g^{\frac{1}{6}} \cdot D_i^{\frac{1}{2}}}{H^{\frac{1}{2}} \cdot D_t^{\frac{3}{2}}} \tag{33}$$

with

t = mixing time [sec]
D_i = impeller diameter [m]
D_t = vessel diameter [m]
H_L = liquid height [m]
g = acceleration due to gravity (m sec^{-2}]
n = rotation speed of impeller [sec^{-1}]

and plotted it against the Reynolds number. Fig. 5 is very similar to the one found for the power factor Np. If again geometrically similar vessels and turbulent flow are assumed, the following expression may be derived from equation (33)

$$tn^{\frac{2}{3}} \cdot D_t^{\frac{1}{6}} = \text{const.}$$

with constant specific power input

$$n^3 D_t^2 = \text{const.}$$

by eliminating n from the two expressions we finally find

$$t_2 = t_1 \left(\frac{D_{t2}}{D_{t1}}\right)^{\frac{11}{18}} \tag{34}$$

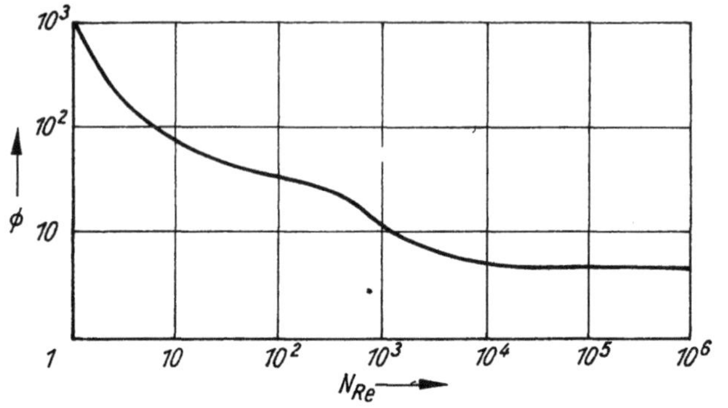

Fig. 5. Mixing time factor versus REYNOLDS number N_{Re}

This means that with our linear scale up factor of 4.64 the mixing time in the vessel is increased by a factor of 2.55, which is quite much. This again makes clear, that scale upe is not unlimited.
Therefor it is worthwhile to see what could be done to minimise t in large fermenters.
If we rearrange (33) we find:

$$t = \phi \frac{H_L^{\frac{1}{2}} \cdot D_t^{\frac{3}{2}}}{(nD_i^2)^{\frac{2}{3}} \cdot g^{\frac{1}{6}} \cdot D_i^{\frac{1}{2}}} \tag{35}$$

we substitute

$$D_i = \beta D_t$$

where β is the ratio $\frac{D_i}{D_t}$ we find

$$t = \frac{\phi}{n^{\frac{2}{3}} g^{\frac{1}{6}}} \cdot \frac{H_L^{\frac{1}{2}}}{\beta^{\frac{11}{6}} D_t^{\frac{1}{3}}} \tag{36}$$

From this expression it becomes clear, that a fermenter with a short mixing time t has rather to have a relatively large impeller a small liquid height and a relatively large tank diameter. This holds even true if we assume that the two vessels (the tall and the shallow one) have impellers with the same tip speed. This fact can be shown as follows.

Since tip speed is proportional to nD_i we may write

$$V_{\text{tip}} \approx n\beta D_t = \text{const.} \qquad n = \frac{\text{const.}}{\beta D_t}$$

and by substituting n in (36) we find

$$t = \frac{\phi}{\text{const}^{\frac{2}{3}} g^{\frac{1}{6}}} \cdot \frac{H_L^{\frac{1}{2}} D_t^{\frac{1}{3}}}{\beta^{\frac{7}{6}}} \tag{37}$$

which makes clear that under the assumptions made, the stirrer diameter is of utmost importance with respect to mixing time. The examples shown make it evident that scale up of a fermenter is very complex. Nevertheless the methods outlined in this short paper are a good guideline. The ideal fermenter does not yet — and certainly never will — exist. It has always to be made to measure to the problem under consideration and always good compromises will have to be made.
This on one side may be disappointing for the ambitious engineer, but on the other hand it is also a challenge to work in a field where at least a part of it is still more an art than a science.

Eingegangen: 12. 12. 1980

References

AIBA, S., HUMPHREY, A. E., MILLIS, N. F., Biochemical engineering, Academic Press, New York, (1965).
AIBA, S., and YAMADA, T.: "Oxygen absorption in bubble aeration. Part 1." J. Gen. Appl. Microbiol. 7 (1961) 100.
CALDERBANK, P. H.: "Physical rate process in industrial fermentation. Part 1. The interfacial area in gas-liquid contacting with mechanical agitation." Trans. Instn. Chem. Engrs. **36** (1958) 443.
CALDERBANK. P. H., and MOO-YOUNG, M. B. (1959), and (1961). "The prediction of power consumption in the agitation of non-Newtonian fluids." ibid. 37, 26. "The power characteristics of agitators for the mixing of NEWTONian and non-NEWTONian fluids." ibid. 39, 337.
NORWOOD, K. W., and METZNER, A. B.: "Flow patterns and mixing rates in agitated vessels." A. I. Ch. E. Journal **6** (1960) 432.
OHYAMA, Y., and ENDOH, K.: "Power characteristics of gas-liquid contacting mixers." Chem. Eng. (Japan) **19** (1955) 2.
RUSHTON, J. H.: "The use of pilot plant mixing data." Chem. Eng. Progress **47** (1951) 485.
RUSHTON, J. H., COSTICH, E. W., and EVERETT, H. J.: "Power characteristics of mixing impellers. Part 2." Chem. Eng. Progress **46** (1950) 467.
RUSHTON, J. H.: Comment presented at the 25th Anniversary of Soc. Chem. Eng. (Japan). Chem. Eng. (Japan) 26 (1962) 958.
NAGATA, S., (1975), "Mixing, Principles and applications." John WILEY & Sons, New York.
ULLMANNS, (1972), "Encyklopädie der technischen Chemie." "Allgemeine Grundlagen der Verfahrens- und Reaktionstechnik." Band 1, Verlag Chemie, Weinheim.

Acta Biotechnologica 2 (1982) 1, 72

BOOK REVIEW

H. H. Weetall, G. P. Royer

Enzyme Engineering, Vol. 5

New York—London, 1980, Plenum Press
491 Seiten; 49.50 $

Diese Buchserie faßt die Fortschrittsberichte, die auf den seit 1971 stattfindenden internationalen Konferenzen über „Enzyme Engineering" gehalten wurden, zusammen; vorliegender Band die der „5. International Enzyme Engineering Conference" in Henniker, N. H., 1979.
Neben dem Einleitungsvortrag von E. Katchalsi-Katzir über den gegenwärtigen Stand und die Tendenzen auf diesem Gebiet werden 79 weitere Vorträge dargeboten, die in 7 Sektionen zu den Themen: Enzymproduktion, Enzyme und Energieübertragung, Biomasseumwandlung, biomedizinische und analytische Anwendungen von Enzymen und Antikörpern, Umwandlungen im größeren Maßstab mit immobilisierten Enzymen, immobilisierte Zellen und Organellen, immobilisierte Enzyme und Synthese von Feinchemikalien sowie Pharmazeutika — gehalten wurden.
In der Mehrzahl der Vorträge wird über immobilisierte Enzyme (38) oder über die immer mehr an Interesse gewinnenden immobilisierten Zellen bzw. Zellorganellen (17) berichtet. 9 Vorträge befassen sich mit der Isolierung, Reinigung und mikrobiellen Produktion von Enzymen.
So wie in den vorhergehenden Bänden dieser Serie lassen sich der aktuelle Stand und die gegenwärtigen Tendenzen auf diesem Forschungs- und z. T. bereits Anwendungsgebiet ablesen, vorgetragen von auf diesem Gebiet international anerkannten Experten und Forschungsgruppen.
Bei der Enzymproduktion dürften in der Zukunft deutliche Fortschritte durch die „genetic engineering"-Methoden zu erwarten sein. In diesem Zusammenhang sind noch einige Probleme der „large scale"-Isolierung und Reinigung dieser Enzyme zu lösen; ebenso bezüglich deren Stabilisierung durch Immobilisierung oder durch andere Manipulationen. Die Testung der Anwendung von mehreren gemeinsam immobilisierten Enzymen (immobilisierte Enzymkaskaden) sowie die der möglichst billigen Coenzymregenerierung (Mehrfachverwendung) steht nach wie vor im Blickpunkt. Hier scheint sich eine partielle Lösung des Problems durch die Verwendung immobilisierter lebender Zellen anzubieten. Die sich abzeichnende weltweite Energiekrise fördert die Suche nach biochemischen bzw. mikrobiellen Alternativlösungen zur Energiegewinnung (Wasserspaltung in H_2 und O_2, Ethanol aus Glucose, Glucose aus Cellulose mittels Cellulasekomplex, u. a.).
Zu diesen Problemen findet der Leser wichtige und interessante Hinweise, basierend auf meist neuesten Forschungsergebnissen. Dieses Buch, wie auch die vorhergehenden Bände dieser Serie, sollte in jeder Forschungs- und Industrieeinrichtung, die sich mit „Enzyme Engenierung" befaßt, vorhanden sein.

R. Berger

Acta Biotechnologica 2 (1982) 1, 73—77

Effektivität von Modellierung und Optimierung mikrobieller Prozesse

M. PRAUSE und K. SOYEZ

Akademie der Wissenschaften der DDR
Institut für technische Chemie Leipzig
7050 Leipzig, Permoserstr. 15

Summary

The aim of the technological treatment of microbial processes is the optimal working of the production plant. The treatment itself is to be seen as a process whose realization requires a certain amount of expenditure. The problem of optimization of this process can with certain restrictions be approximately solved in a series of steps. After a general formulation of the problem explanations are given for two important and typical steps — the determination of the extend of modelling and the choice of production cultures.

Zusammenfassung

Ziel der technologischen Bearbeitung von mikrobiellen Prozessen ist die optimal arbeitende Produktionsanlage. Die Bearbeitung selbst ist als ein Prozeß anzusehen, für dessen Realisierung gesellschaftliche Mittel aufgewendet werden müssen. Das Problem der optimalen Gestaltung dieses Prozesses ist unter gewissen Einschränkungen näherungsweise in Teilschritten lösbar. Nach einer allgemeinen Formulierung des Problems werden Erläuterungen zu zwei relevanten Teilschritten dieses Prozesses gegeben — der Auswahl von Produktionskulturen und der Festlegung des Umfanges von Modellierungsarbeiten.

Ziel der technologischen Bearbeitung von Prozessen auf der Basis von biologisch vorgebildeten Wirkprinzipien ist die Auslegung von Produktionsanlagen und die Führung der darin ablaufenden Prozesse in der Weise, daß eine maximale Effektivität erreicht wird. Dabei hat auch in der Biotechnologie die Bearbeitungsmethode

$$\text{Systemanalyse} \rightarrow \text{Modellbildung} \rightarrow \begin{cases} \text{optimale Anlagenauslegung} \\ \text{optimale Prozeßführung} \end{cases} \tag{1}$$

ihren festen Platz gefunden. Es ist in diesem Entwicklungsstadium angezeigt, daß der Biotechnologe seine Aufmerksamkeit verstärkt einem Problem zuwendet, das mit der Effektivität dieser Bearbeitungsmethode selbst im Zusammenhang steht (siehe auch [1, 2]).

Das Optimierungsproblem der Bearbeitungskette (1) kann allgemein wie folgt ausge drückt werden:

$$Z_1 = Z_1(\boldsymbol{K}', \boldsymbol{X}, \boldsymbol{A}) \rightarrow \underset{\boldsymbol{X}}{\text{Min}} \tag{2}$$

Z_1 gesellschaftlich relevante Zielgröße
K' Vektor der Koeffizienten von Prozeßmodellen
X Vektor der Steuervariablen
A Vektor von Parametern aus dem gesellschaftlichen Bereich

Die Lösung dieses Optimierungsproblems erfordert gesellschaftliche Aufwendungen, z. B. für die Konkretisierung des Optimierungsproblems durch Erarbeitung von Prozeßmodellen. Da im allgemeinen bei Erhöhung des Bearbeitungsaufwandes ein verbessertes Ergebnis in Z_1^*, dem Optimum von Z_1, erreicht werden kann, wird der Bearbeitungsaufwand selbst zu einer Optimierungsgröße, und das im technologischen Sinne vollständige Optimierungsproblem kann angegeben werden zu:

$$Z_2 = Z_2\left(X, K'(C), A, C\right) \to \min_{\substack{X \in U(C) \\ C}} \qquad (3)$$

C Parametervektor des Bearbeitungsaufwandes
U Raum der Steuerparameter

Der Vektor C wird Bestandteil des Steuervektors. Bezieht man zunächst die Aspekte Parameterungenauigkeit und -unsicherheit nicht in die Optimierungsproblematik ein, dann beschreiben $K' = K'(C)$ und $U = U(C)$ die oben erwähnte Möglichkeit, durch steigenden Bearbeitungsaufwand den Prozeß umfassender zu beschreiben und Z_1^* zu verbessern. Eine Aufgabe (3) konkret zu formulieren und zu lösen ist in dieser umfassenden Form wahrscheinlich auch in Zukunft wegen ihrer Komplexität nicht sinnvoll. Wesentlich ist:

- Festzuhalten, daß Teilschritte der Bearbeitungskette (2) auf ihre Effektivität nur unter Kontrolle ihrer Auswirkungen auf das Gesamtziel Z_2 überhaupt beurteilbar werden
- Nach Methoden zu suchen, die unter Verzicht auf die komplette Lösung von Aufgabe (3) eine hinreichend gute Näherung gestatten.

Eine solche Methode sehen wir in folgendem: Sind — wie in den meisten praktischen Fällen — Bearbeitungszeiten, Kapazitäten und Mittel für Forschung und Entwicklung vorgegeben und kann vorausgesetzt werden, daß der in dieser Weise fixierte Aufwand den nach Aufgabe (3) optimalen nicht übersteigt:

$$F(C) \leqq F^*(C)^*; \quad (C)^* = \left(C \mid Z_2 = Z_2^*(X^*, K'(C)^*, A, C^*)\right), \qquad (4)$$

F Abbildung des Parametervektors des Bearbeitungsaufwandes in den Kostenbereich

so kommt es offensichtlich darauf an, aus der Vielzahl der möglichen Bearbeitungsaktivitäten diejenigen auszuwählen, die nach Aufgabe (2) die besten Resultate in Z_1^* versprechen. Je weniger sich diesbezügliche konkrete Ergebnisse verallgemeinern lassen, um so wesentlicher ist es, den Prozeß von Forschung und Entwicklung in Echtzeit zu betrachten und einen auf die Zielgröße Z_2 gerichteten Regelungseffekt dadurch zu erreichen, daß die notwendigen Entscheidungen zu einem frühen Zeitpunkt — entsprechend geringer Information und großem Risiko bei noch geringen verausgabten Aufwendungen — vorgenommen und mit fortschreitender Realisierungszeit sukzessive korrigiert werden.

In der biotechnologischen Praxis betreffen solche Entscheidungen vor allem folgende Situationen:

- Es liegen zur Bearbeitung mehrere diskrete Varianten vor, z. B. in Form unterschiedlicher Apparate, unterschiedlicher Apparateschaltungen, verschiedener potentieller Produktionskulturen
- Für eine fixierte Verfahrensstufe muß die Frage beantwortet werden, für welche Prozeßeigenschaften Modelle aufgebaut und einer Optimierung zugeführt werden sollen.

Für beide Situationen soll die vorgeschlagene Methodik an je einem Beispiel — der Bewertung von Produktionskulturen und der Wichtung von Modellierungsarbeiten — erläutert werden.
Vorausgeschickt werden muß, daß, wie bei jeder Optimierungsaufgabe, deren Lösung aus einem einzigen optimalen Steuervektor bestehen soll, aus formalen Gründen für Z_1 eine skalare Präferenzfunktion zu fordern ist. Unter dieser Voraussetzung gelten die folgenden Erläuterungen allgemein. Darüber hinaus soll die erhaltene Lösung inhaltlich relevanten gesellschaftlichen Zielstellungen möglichst gut entsprechen. Eine Präferenzfunktion, die beide Forderungen hinreichend gut erfüllt, ist eine spezifische Aufwandskennziffer [3], die in allgemeiner Form angegeben werden kann als:

$$aw = \frac{1}{A} (I \cdot c + U \cdot d + b) \quad [\mathrm{M/t}] \tag{5}$$

I Investkosten [M]
U Umlaufmittel [M]
b Betriebsaufwand [M/a]
A jährlicher Ausstoß [t/a]
c, d ökonomische Konstanten [1/a]

Diese spezifische Aufwandskennziffer ist als Optimierungskriterium nur anwendbar für den Fall eines konstanten Gebrauchswertes des Endproduktes.
Zum Zwecke der Bewertung von Produktionskulturen ist es notwendig, innerhalb des Vektors von Koeffizienten des Prozeßmodells einen Vektor von Kultureigenschaften, $\boldsymbol{K}$, getrennt auszuweisen:

$$\boldsymbol{K}' = (\boldsymbol{K}, \boldsymbol{K}'')^T . \tag{6}$$

In Übereinstimmung mit (2) und (6) wird die Güte eines mikrobiell-technischen Verfahrens von den Eigenschaften der Produktionskultur maßgeblich mitbestimmt.
Der Bereitstellung geeigneter Produktionskulturen ist deshalb ein nicht unbeträchtlicher Teil der Forschungskapazität zugeordnet. Als Folge dieser Maßnahme liegen zu einem bestimmten Zeitpunkt eine Anzahl potentieller Produktionskulturen vor. Es ergeben sich die Aufgabenstellungen

- nach Wegen der Aufwandsreduzierung bei der Gewinnung von Produktionskulturen zu suchen
- in einem frühen Stadium die Menge der weiter zu bearbeitenden Kulturen auf diejenige einzuschränken, die für das Verfahren das beste Resultat verspricht.

Insbesondere die letztere Forderung ergibt sich daraus, daß die erforderlichen Arbeiten zur Auslegung von Verfahrensstufen (im Idealfall die Gewinnung von Modellen, der Aufbau von Zielfunktionen und deren Optimierung) und die Untersuchungen zur Bestätigung der medizinisch-biologischen Unbedenklichkeit aufwendig (speziell zeitaufwendig) sind. Beide Aufgabenstellungen hängen auf das engste zusammen mit einer Bewertung der potentiellen Produktionskulturen als Voraussetzung für die zu treffenden Entscheidungen.
Diese Bewertung kann nur unter Berücksichtigung quantifizierbarer Kultureigenschaften erfolgen. Dabei ergibt sich das Problem, daß unterschiedliche Kulturen mit jeweils mehreren Merkmalen nur im Sinne der Halbordnung von Vektoren [4] miteinander verglichen werden können. Eine formal eindeutige und inhaltlich sinnvolle Bewertung gelingt über das Kriterium

$$Z_1^* = Z_1^*\big(\boldsymbol{K}, \boldsymbol{K}'', \boldsymbol{X}^*(\boldsymbol{K})\big) . \tag{7}$$

Soll eine gegebene Kultur auf der Basis von Z_1^* beurteilt werden, so ist mit dem gegebenen Vektor $\boldsymbol{K}$ eine Optimierungsrechnung

$$Z_1^* = \operatorname*{Min}_{\boldsymbol{X}} Z_1 \tag{8}$$

durchzuführen. Eine Vergleichsmöglichkeit von n Kulturen ist über den Vergleich von Z_{1j}^*, $j = 1, \ldots, n$, gegeben. Wird ein solcher Vergleich häufig gefordert, so ist es angezeigt, das Entscheidungskriterium bequemer handhabbar zu machen. Dazu bietet sich an, Z_1^* als Funktion von $\boldsymbol{K}$ aufzufassen und in einem einfachen Ausdruck näherungsweise darzustellen. Untersuchungen zur Fermentationsstufe kontinuierlicher Verfahren zur Erzeugung mikrobieller Biomasse aus nichtkonventionellen C-Quellen haben gezeigt, daß eine solche Darstellung mit dem Vektor

$$\boldsymbol{K} = (\alpha_x^{O_2}, \alpha_x^{CO_2}, \alpha_x^{Q}, (\alpha_x^{NS})_1 \ldots (\alpha_x^{NS})_l, \vartheta)^T \tag{9}$$

α_x Verbrauchs- bzw. Bildungskoeffizienten
Q Wärmemenge
C Kohlenstoffsubstrat
NS Nährsalz
ϑ Reaktionstemperatur

als Polynom 2. Ordnung möglich ist:

$$\begin{aligned} Z_1^* = a_0 &+ a_1\alpha_x^{C} + a_2\alpha_x^{O_2} + a_3\alpha_x^{CO_2} + a_4 \sum_{i=1}^{l} c_i(\alpha_x^{NS})_i \\ &+ a_5\alpha_x^{Q} + a_6\vartheta \\ &+ a_7(\alpha_x^{O_2})^2 + a_8(\alpha_x^{CO_2})^2 + a_9(\alpha_x^{Q})^2 + a_{10}\vartheta^2 \\ &+ a_{11}\alpha_x^{O_2}\alpha_x^{CO_2} + a_{12}\alpha_x^{O_2}\alpha_x^{Q} + a_{13}\alpha_x^{O_2}\vartheta \\ &+ a_{14}\alpha_x^{CO_2}\alpha_x^{Q} + a_{15}\alpha_x^{CO_2}\vartheta + a_{16}\alpha_x^{Q}\vartheta. \end{aligned} \tag{10}$$

Eine derartige Darstellung macht die Ermittlung von $Z_{1,j}^*$ für beliebige Kulturen äußerst einfach. Deshalb sind auch Erweiterungen des Anwendungsbereiches im Bereich des Möglichen. Gleichung (10) ist zunächst nur als Entscheidungsmodell gedacht, d. h., die zu beurteilenden Kulturen sind vorhanden, die Elemente des Vektors $\boldsymbol{K}$ mehr oder weniger gut bekannt. Modelle des Typs (10) können aber auch eine Rolle im Prozeß der Selektion von Kulturen als Steuermodell spielen, wenn es gelingt, Elemente eines Vektors $\boldsymbol{K}$ als Steuervariable für den Selektionsprozeß einzuführen.

Das vorgeschlagene Bewertungsmodell (7) in seiner allgemeinen Form ist prinzipiell auf jeden biotechnischen Prozeß der Biomasseproduktion anwendbar. Die resultierende Aussage ist entsprechend den eingebrachten Voraussetzungen nicht vollständig und mit Risiko behaftet. Der Vorzug der Methode besteht darin, im Bereich der quantifizierten Kultureigenschaften eine weitgehend objektive Bewertung zu ermöglichen. Die Forderung nach weiterer Verbesserung deckt sich mit dem Bestreben des Biotechnologen, auch auf dem Gebiet der optimalen Auslegung Modelle und Zielfunktionen weiter aus- und Risiken schrittweise abzubauen.

Zur Beantwortung der Frage, welche Prozeßeigenschaften modelliert werden sollen, kann zunächst festgestellt werden, daß Optimierungen mit verbesserten, d. h. um den Steuervektor $\boldsymbol{v}$ erweiterten Prozeßmodellen dadurch sinnvoll sein können, daß sie gegenüber Optimierungen unter Benutzung von Modellen ohne diese Erweiterung stets gleiche oder bessere Resultate in der optimalen Zielgröße liefern:

$$Z_1^*(\boldsymbol{X}^*) = \operatorname*{Min}_{\boldsymbol{X}}\{Z_1(\boldsymbol{X})\} \geqq \operatorname*{Min}_{\boldsymbol{X},\boldsymbol{v}} \{Z_1'(\boldsymbol{X}, \boldsymbol{v})\} = Z_1'^*(\boldsymbol{X}^*, \boldsymbol{v}^*) \tag{11}$$

Zu fragen ist, ob diese Verbesserungen in einem sinnvollen Verhältnis zum getriebenen Aufwand stehen. Eine Antwort liefert die Kosten-Nutzen-Analyse. Es wurde bereits vorausgesetzt, daß diese Frage positiv beantwortet sein soll. Es handelt sich in diesem Fall darum, bei gegebenem Aufwand die erfolgversprechendsten Modellierungsarbeiten auszuwählen. Auch dazu ist eine Abschätzung des zu erwartenden Nutzens Voraussetzung. Diese Abschätzung kann über einen Vergleich von Z_1^* und $Z_1'^*$ erfolgen.
Von den zur Modellierung in Betracht gezogenen Prozeßeigenschaften muß, wenn ein nichttriviales Optimierungsergebnis resultieren soll, gefordert werden, daß sie eine Kompromißsituation in Z_1 aufbauen können. Als konkretes Beispiel sei folgende für den stationären Fermentationsprozeß typische Kompromißsituation genannt: Steigende Reaktionstemperatur bewirkt einerseits eine Senkung von Z_1 auf Grund verringerter Aufwendungen für die Reaktorkühlung, andererseits aber eine Erhöhung von Z_1 auf Grund der im Bereich $\vartheta > \vartheta^\circ$,

$$\vartheta^\circ = (\vartheta \mid \alpha_x{}^i = \alpha^i_{x,\min} := \alpha_x{}^{i_0}), \tag{12}$$

mit ϑ ansteigenden stöchiometrischen Koeffizienten $\alpha_x{}^i$.
Welchen Nutzen würde eine Modellierung für $\alpha_x{}^i = \alpha_x{}^i(\vartheta)$ erbringen können?
Zur Beantwortung dieser Frage ist $Z_1^{*\prime}$ durch Optimierungsrechnungen mit einer vorläufigen Information in Form eines zunächst groben Modells $\alpha_x{}^i = \alpha_x{}^i(\vartheta)$ zu ermitteln. Besteht die vorläufige Information aus nur so wenigen zusammenhängenden Wertepaaren $\alpha_x{}^i$ und ϑ, daß ein Modell schwer ableitbar ist, dann können Optimierungsrechnungen für $Z_1^*(\vartheta)$ erste Anhaltspunkte über $Z_1'^*$ ergeben. In jedem Fall ist die Ermittlung von $Z_1'^*$ aufwendig und wird wesentlich erleichtert, wenn auf ein Bewertungsmodell des Typs (10) zurückgegriffen werden kann. Dann ist $Z_1^*(\vartheta)$ aus Gleichung (10) für zusammengehörige Wertesätze $\alpha_x{}^i$ und ϑ mühelos zu berechnen und eine gute Information über $Z_1'^*$ zu gewinnen. Ist etwa die Temperaturabhängigkeit eines stöchiometrischen Koeffizienten, z. B. $\alpha_x{}^{O_2}$, besser bekannt als die der übrigen, ist eine Näherung für diese unter Verwendung von

$$\alpha_x{}^i = \alpha_v{}^i \cdot \alpha_x{}^{O_2}$$

$\alpha_v{}^i$ Verknüpfungskoeffizienten

zu empfehlen. Auf diese Weise wurden Abschätzungen mit Literaturdaten für verschiedene aerobe Fermentationen nichtkonventioneller C-Substrate durchgeführt und ergaben zu erwartende Optimierungseffekte von $\sim 5\%$ von Z_1^*. Die auf Grund dieser Aussage durchgeführten Modellierungs- und Optimierungsarbeiten, deren schrittweise anfallende Ergebnisse zur Präzisierung der anfänglichen Aussagen verwendet wurden, konnten je nach den konkreten verfahrenstechnischen und ökonomischen Bedingungen einen Optimierungseffekt von 2—3% von Z_1^* mit $\vartheta^* > \vartheta^\circ$ nachweisen. Das ist in dieser Größenordnung ein Effekt, der im allgemeinen als durchaus erfolgreich bei technologischen Optimierungen angesehen werden kann.

Eingegangen: 10. 4. 1981

Literatur

[1] Soyez, K., Prause, M., „Eine Methode zur Bewertung von Genauigkeiten von Meßgrößen in der Biotechnologie“ 2. Symposium der sozialistischen Länder über Biotechnologie, Leipzig 1980, Poster-Sektion 3.
[2] Soyez, K., Theiss, D., Prause, M., „Technologische Bewertung von Meßanordnungen und Meßgenauigkeiten“ Abhandlungen der AdW der DDR, N 3 (1979), 299.
[3] Hildebrand, H.-J., Hedrich, P., Ufer, D., „Wirtschaftlichkeitsrechnung“, VEB Verlag für Grundstoffindustrie, Leipzig 1970.
[4] Peschel, M., Riedel, C., „Polyoptimierung“, VEB Verlag Technik, Berlin 1976.

Acta Biotechnologica 2 (1982) 1, 78

BOOK REVIEW

BERND KÖGL; FRANZ MOSER

Grundlagen der Verfahrenstechnik

Springer-Verlag Wien/New York 1981
377 Seiten 138 DM

Dieses Buch beinhaltet die Kapitel Grundbeziehungen der Austauscheffekte, die Bilanzgleichungen, Ähnlichkeitstheorie, Analogie der Austauscheffekte, Stoffaustausch in einphasigen Systemen, Stoffübergang an testen Grenzflächen, Stoffaustausch durch Grenzflächen, Stoffübergang am Flüssigkeitsfilm sowie an Blasen und Tropfen und Stoffaustausch und chemische Reaktion.
Das Buch geht aus Vorlesungen zur Verfahrenstechnik hervor, die über viele Jahre an der TU Graz gehalten wurden. Das Anliegen der Autoren war dabei, die wesentlichen Grundlagen der Verfahrenstechnik in systematischer Form darzustellen, um insbesondere dem Studierenden ein Buch in die Hand zu geben, in dem dieses Wissensgebiet für ihn geschlossen und gut überschaubar vorliegt.
Im Vergleich zu anderen deutschsprachigen Lehrbüchern der Verfahrenstechnik, wie Grassmann „Physikalische Grundlagen der Verfahrenstechnik" oder Kassatkin, Adolphi, Vauck, Müller u. a. unterscheidet sich das vorliegende Buch in Aufbau und Darlegung des Stoffes sehr wesentlich. Während erstere vorzugsweise von den verfahrenstechnischen Grundoperationen wie Destillation, Extraktion, mechanische Stofftrennung usw. ausgehen und damit dem Praktiker entgegenkommen, stellen im Gegensatz dazu Kögl und Moser die theoretischen Grundlagen für diese Prozesse in den Mittelpunkt, um damit für den Studierenden das theoretische Rüstzeug zu schaffen, das für die Behandlung dieser Grundoperationen erforderlich ist. Die Zahl der praktischen Beispiele ist deshalb auch knapp bemessen, ausreichend zur Erläuterung der Theorie, aber sicherlich erschwerend für denjenigen, der die Absicht hat, nur bestimmte Abschnitte zu lesen.
Deutlich sind solche Gebiete wie Ähnlichkeitstheorie und Stoffaustausch in den Mittelpunkt gestellt, die für den in Forschung und Entwicklung tätigen Verfahrensingenieur von besonderer Bedeutung sind. Als sehr wichtig wird in dieser Hinsicht empfunden, daß bei der Behandlung der Ähnlichkeitstheorie besondere Beachtung der Prüfung der Anwendbarkeit und der Beschreibung der Leistungsgrenzen der Ähnlichkeitstheorie geschenkt wird.
Für ein weitergehendes Studium befinden sich am Anfang und Ende jeden Kapitels ausreichende Literaturhinweise.

W. STEPHAN

Acta Biotechnologica 1 (1982) 1, 79–85

Regulation of Lysine Biosynthesis in Brevibacterium flavum

M. P. Leite, M. P. Rukliša, U. E. Viesturs, J. E. Švinka

August Kirchenstein Institute of Microbiology
Latvian SSR Academy of Sciences, Riga, USSR

Summary

L-lysine synthesis pathway enzyme activities: β-aspartate kinase (EC. 2.7.2.4), diaminopimelate decarboxylase (EC.4.1.1.20) for two L-lysine producing strains *Brevibacterium flavum* 22LD and RC-115 were studied. It has been found that β-aspartate kinase and diaminopimelate decarboxylase in the *Br. flavum* RC-115 are less sensitive to feed-back inhibition by lysine and threonine. It is supposed that desensitized β-aspartate kinase in the *Br. flavum* RC-115 can be determined by genetical changes of the regulatory properties of the β-aspartate kinase.
Auxotrophity in the locus of homoserine dehydrogenase was tested and no homoserine dehydrogenase (EC.1.1.1.3) activity was found in either strain.
The combination of these both types of mutation supplemented by the lack of catabolic repression in the RC-115 strain makes it an active lysine producer in the medium with high carbohydrates content.

Zusammenfassung

Die Aktivität der Enzyme des L-Lysin-Syntheseweges von zwei L-Lysin-produzierenden Bakterienstämmen *Brevibacterium flavum* 22LD und *Brevibacterium flavum* RC-115 wurde untersucht: der β-Aspartatkinase (EC 2.7.2.4.) und der Diaminopimelatdecarboxylase (EC 4.1.1.20.). Es wurde festgestellt, daß die β-Aspartatkinaseaktivität des RC-115-Stammes durch die Rückkopplung mit L-Lysin und L-Threonin weniger inhibiert wird. Vermutlich ist die verminderte Sensibilität der β-Aspartatkinase des RC-115-Stammes durch die veränderte Enzymregulation genetisch determiniert. Es wurde festgestellt, daß beide Mutantenstämme keine Homoserindehydrogenaseaktivität (EC 1.1.1.3.) besitzen. Die vorhandenen Mutationen in Kombination mit dem Ausfall der Katabolitrepression bei *Brevibacterium flavum* RC-115 ergeben eine hohe Aktivität der L-Lysin-Synthese des Stammes.

Introduction

Intensive technology of microbiological production of L-lysine is based on the employment of active producers. For microbiological synthesis of L-lysine mutants are used which lack certain enzymes or the enzymes regulatory properties are changed. It will be possible to implement the potential lysine synthesizing properties of these mutants when the producer metabolism and its regulatory principles are completely researched.
The economy of microbiological synthesis is to a great degree defined by the carbon metabolism of producers. Lysine biosynthesis is directly connected with the metabolism

of carbohydrates, as well as with the formation of homoserine, threonine, methionine and isoleucine — compounds which like lysine are aspartic acid family amino acids. Earlier the study of carbon metabolism enabled us to define 4 alternative lysine biosynthesis pathways in *Brevibacterium flavum* 22LD [1]. The regulation of lysine synthesis can be performed by the control of carbohydrate metabolism intensity via glycolysis and the tricarbonic acid (TCA) cycle enzymes [2]. The metabolic way of improving lysine biosynthesis includes induction of the TCA cycle and the glyoxylate cycle enzymes [2, 3, 4], maintaining high phosphoenol-pyruvate carboxylase activity [1], eliminating of by-product synthesis [3, 4, 5]. The technological approaches for obtaining high carbon conversion yields $\left(Y_p = \frac{\text{g lysine}}{\text{g substrate}}\right)$ with *Br. flavum* are discussed in [5, 6, 7].

The other condition for obtaining high lysine yield is limitation of feedback inhibition of β-aspartate kinase [8] (the first step enzyme for lysine synthesis from aspartate) and diaminopimelate (DAP) decarboxylase (the last enzyme in lysine biosynthesis pathway) [9].

The β-aspartate kinase regulatory control differs in cell extracts of *Corynebacterium glutamicum* [10] and *Br. flavum* [11, 12] parent and mutant strains. The genetically determined difference of allosteric properties of β-aspartate kinase was found in homoserine and threonine auxotrophic mutants and S(2-aminoethyl)-L-cysteine (AEC) resistant mutants [10]. The decreased sensitivity of β-aspartate kinase to feedback inhibition by L-lysine plus L-threonine of AEC-resistant mutants favoured the increased accumulation of lysine.

The other possibility for lysine overproduction is a genetically defected homoserine dehydrogenase [13]. Combination of both types of mutation seems to be the best way to increase the production of lysine.

The present investigation is focussed on the study of homoserine dehydrogenase (EC. 1.1.1.3), β-aspartate kinase (EC.2.7.2.4), diaminopimelate decarboxylase (EC.4.1.1.20) activities in two lysine producing *Br. flavum* strains: 22LD and RC-115 in order to choose the producer with higher potential activity.

Materials and methods

Microorganisms and Cultivation: The lysine-producing strains of *Brevibacterium flavum* 22LD [14] and RC-115 [15] were studied.

Fermentation was carried out in laboratory-type fermenters FS-5A [16] at temperature 30 °C and pH 7.5—7.8. The aeration was controlled according to the partial pressure of dissolved oxygen (pO_2) as described in [3].

The inoculum of the culture was grown for 18 hours at 30 °C in Erlenmeyer flask containing 50 ml seed medium with shaking. The composition of the seed medium was as follows:

molasses — 2% (to the reducing substances)
corn steep extract — 2.5% (dry weight)
$(NH_4)_2SO_4$ — 1%

The inoculum (1.5 g/liter) was transferred into the fermentation medium. The standard fermentation medium contained:

molasses — 7.5% (to the reducing substance)
corn steep extract — 2% (dry weight)
$(NH_4)_2SO_4$ — 2%
KH_2PO_4 — 0.05% and K_2HPO_4 — 0.05%.

Sometimes for obtaining the enzyme preparations the bacterial mass was cultured in synthetic fermentation medium containing:

sucrose — 7%
$(NH_4)_2SO_4$ — 2%
K_2HPO_4 and KH_2PO_4 — 0.05% each
$MgSO_4 \cdot 7H_2O$ — 0.04%, $FeSO_4 \cdot 7H_2O$ — 0.001%
$MnSO_4 \cdot 4H_2O$ — 0.0005%
biotin — 3.0 μg %,
thiamine — 15.0 μg %
L-methionine — 10 mg %
L-threonine — 40 mg %
$CaCO_3$ — 2%.

Preparation of enzyme solution

The bacterial mass was separated from $CaCO_3$ by centrifugation 10 min at 3300 g, then suspended in 0.9% NaCl and once washed. The cells were harvested at 4 °C and stored at −20 °C.
For obtaining the enzyme preparation the bacterial mass was suspended in 0.1 M Tris-HCl buffer (pH 7.5) + 0.02 M β-mercaptoethanol. The cells were disrupted in a desintegrator (L-17) at 1000 rpm for 3 minutes at +4 °C [6].
After centrifugation at 15000 g for 30 min to the supernatant were added 5 volumes of saturated $(NH_4)_2SO_4$ solution. After 16—18 hours the suspension was centrifugated at 15000 g for 30 min and the precipitate was dissolved in the same buffer (0.1 M Tris-HCl + 0.02 M β-mercaptoethanol; pH — 7.5). All enzyme preparations were stored at −20 °C until use. Immediately before the experiment the enzyme preparations were diluted with water so that the protein contents in the solution would be 13—16 mg/ml.

Determination of enzyme activities

β-aspartate kinase activity was measured by the method of Black and Weight [17] with some modifications. 10 mM Tris-maleic acid — NaOH buffer (pH 7.0) was used. All the components of the standard reaction mixture, except $MgCl_2$ were adjusted to neutrality with KOH prior to addition to the reaction mixture. After incubation for 30 min at 37 °C the reaction was stopped by adding 1.5 ml of $FeCl_3$ reagent [18]. The precipitated protein was removed by centrifugation for 30 min at 15000 g.
The absorbance at 540 nm was registered with spectrophotometer SF-16 (USSR). The amount of aspartate hydroxamate was established according to a standard calibrating curve. The units of enzyme activity were expressed as μmols of aspartate hydroxamate produced on 1 mg enzyme protein per sec.
Diaminopimelate decarboxylase activity was evaluated according to the method described in [18].
The activity of homoserine dehydrogenase was determined in cell extracts. The desintegration of cells and determination of the enzyme activity was carried out in 0.1 M Tris HCl buffer, pH — 9.25 containing 0.8 M KCl. The enzyme activity was evaluated as the NADP reduction rate [19] in the presence of homoserine and registered on Specord UV-VIS (GDR) at 340 nm, at 30°C. The enzyme activity was expressed in μmol/mg protein · min.

Protein determination

Protein concentration was measured by method of Lowry et al. [20].

Chemicals

The amino acids used in these experiments were all in L-form. The ATP and amino acids were purchased from „Reanal" (Hungary), β-aspartate hydroxamate — from "Sigma" (USA). All other chemicals were of reagent grade.

Results and discussion

From the obtained results it follows that the strains *Br. flavum* 22LD and RC-115 have a complete genetic block in the homoserine locus. Homoserine dehydrogenase activity has not been observed in either strain. For comparison the activity of homoserine dehydrogenase of the parent strain *Br. flavum* is $7.7 \cdot 10^{-5}$ μmol/mg · protein · min.

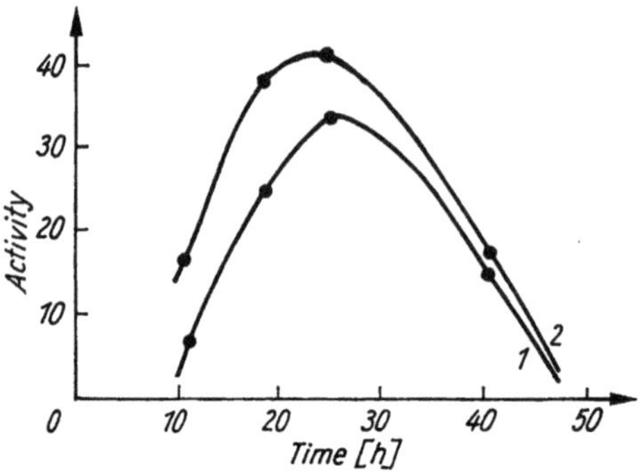

Fig. 1. Aspartate kinase activity of the *Br. flavum* strains 22LD [1] and RC-115 [2], at cultivation in standard molasses media.
Abscissa — cultivation time (hours);
Ordinate — aspartate kinase activity (μmol substrate/mg of protein · min).

The activities of aspartate kinase and DPA decarboxylase of the strains to be compared have been shown in Figures 1, 2. Both producers differed little regarding the aspartate kinase and DAP decarboxylase activities under standard conditions of testing in the absence of amino acids. At cultivation in complex media containing at the beginning of

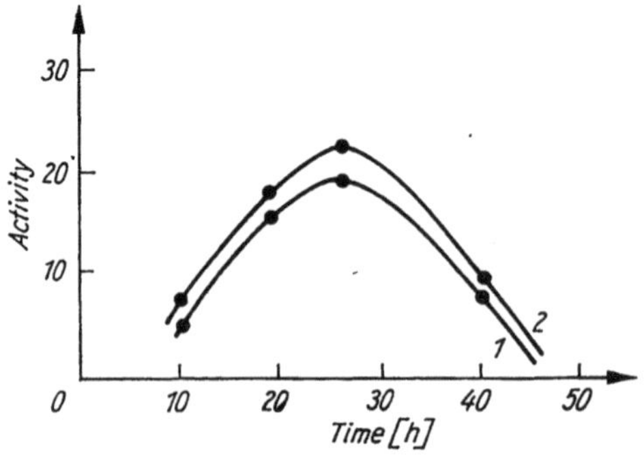

Fig. 2. DAP decarboxylase activity of the *Br. flavum* strains 22LD [1] and RC-115 [2], at cultivation in standard molasses media.
Abscissa — cultivation time (hours);
Ordinate — DAP decarboxylase activity (μmol substrate/mg of protein · min).

the process apart from 40 mg% of threonine also lysine (about 150 mg%) introduced with the molasses and corn steep extract, the aspartate kinase and DAP decarboxylase activity with the RC-115 strain is higher in comparison with the 22LD strain.
Apparently the aspartate kinase and DAP decarboxylase in RC-115 strain are less sensitive to feedback inhibition by lysine plus threonine.
As shown before [21] overproduction of lysine, especially in media with high concentration of reducing substances (RS) with RC-115 strains is conditioned by the absence of catabolite repression of the TCA cycle enzymes. The enhanced metabolism activity via the TCA cycle is evidently also linked with the increased activity of enzymes in the lysine synthesis pathway when lysine and threonine are present in the medium.
Lysine yield from carbohydrates according to our data and those obtained by other authors is equal to 0.27—0.35 [22, 23].

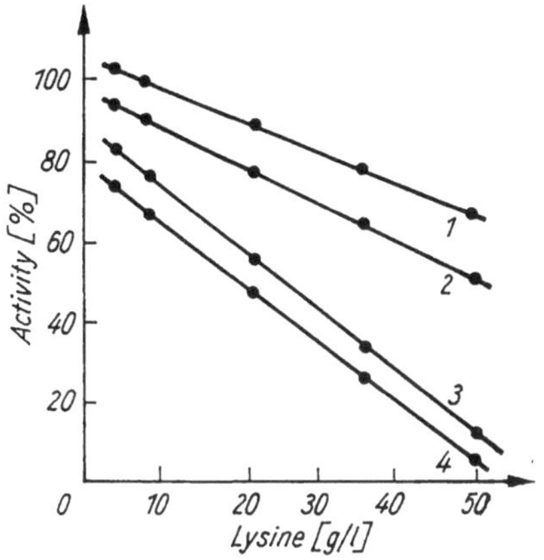

Fig. 3. Influence of various concentrations of lysine and fixed concentration of threonine (0.19 g/l) on the activity of aspartate kinase and DAP decarboxylase.
Abscissa — L-lysine concentration (g/l)
Ordinate — enzyme activity (%)
(100% is the assumed activity of the enzyme at cultivation in the absence of amino acids).
DAP decarboxylase activity: 1 — strain RC-115, 2 — strain 22LD;
Aspartate kinase activity: 3 — strain RC-115; 4 — strain 22LD.

To obtain 50—60 g/liter lysine from molasses medium the RS contents in the medium should be brought up to 16.0—20.0%. However, earlier experiments [2] show that the increase of RS concentration above 12—15% leads to metabolism inhibition via tricarbonic acids cycle. To reduce the inhibitory action of the substrate a fractioned supply of the carbon source — molasses is used [1]. The lysine synthesizing activity is thus increased 2 fold. However, at a fractioned supply of the molasses the concentration of threonine in the medium increases which, together with lysine, suppresses the activity of β-aspartate kinase, and alongside reduces lysine yield [21] (Figures 3, [4]).
Therefore it was of interest of study the activity of aspartate kinase and DAP decarboxylase in the presence of large concentrations of lysine plus small threonine concentrations, as is the case at fractional supply of molasses for increasing the concentration of RS in the medium during fermentation (Figure 3).
Large lysine concentration plus small concentrations of threonine significantly inhibited the aspartate kinase activity of both strains. However, the residual activity of the aspartate kinase in the presence of lysine and threonine mixture was larger for RC-115

than that of the 22LD strain. The activity of DAP decarboxylase was inhibited similarly, but the influence of the lysine and threonine mixture on the activity of this enzyme was less pronounced.

Table 1 presents data about the influence of various proportions of lysine and threonine on the activity of aspartate kinase. At simultaneous addition of equimolar concentrations of lysine and threonine the aspartate kinase activity for the 22LD strain decreased about two times more than the aspartate kinase activity with the RC-115 strain. The influence of little threonine plus much lysine or the influence of little lysine plus much threonine was practically alike — the aspartate kinase activity of the 22LD strain was more inhibited. Both, for the 22LD strain and that of RC-115 the increase of lysine concentration plus small threonine concentration influenced the aspartate kinase activity more than the increase of threonine concentration plus small lysine concentration.

Table 1. Change of aspartate kinase activity in the presence of various lysine and threonine concentrations

L-amino acids	Aspartate kinase activity, %	
	22LD strain	RC-115 strain
without amino acids	100	100
lysine — 20 mM	76	84
threonine — 20 mM	88	93
2mM lysine + 20 mM threonine	48	82
20 mM lysine + 2 mM threonine	36	71
20 mM lysine + 20 mM threonine	28	69

Threonine alone practically did not change the aspartate kinase activity of the RC-115 strain and slightly inhibited the aspartate kinase activity of the 22LD strain. Lysine alone inhibited the aspartate kinase activity at about 20% in both strains. The residual activity of aspartate kinase in the presence of 20 mM of lysine is equal to 76%, (22LD strain) and 84% (RC-115 strain).

According to literature data [24] increased accumulation of lysine by threonine or homoserine auxotrophs may be explained by desensitized aspartate kinase from feedback inhibition by lysine and threonine. The decreasing of enzyme sensitivity to the joint action of lysine and threonine is similar, as is the case with AEC-resistant mutants [10, 11].

The present research shows that aspartate kinase and less pronouncedly DAP decarboxylase of the RC-115 strain has a decreased sensitivity against threonine plus lysine. This property enables to liken the RC-115 strain to the regulatory mutants described in literature with desensibilized aspartate kinase [10].

It has been found that both the strains lack homoserine dehydrogenase. The strain RC-115 differs from the 22LD one with additional mutation in aspartate kinase locus. These properties along with the previously noted absence of catabolite repression enable to use the RC-115 strain as an efficient lysine producer especially for media containing high concentration of carbohydrates. By using the RC-115 strain in pilot equipment 79 g/liter lysine — HCl has been obtained. These results have been obtained by the authors of this article together with M. Beker, G. Liepiņš, A. Sakse, M. Dunce, S. Selga.

The regulatory properties of aspartate kinase and DAP decarboxylase serve as one of the criteria when selecting the producer strains for intensive technology of L-lysine production.

Eingegangen: 5. 3. 1981

References

[1] ŠVINKA, J., VIESTURS, U. and RUKLIŠA, M.: Biotechnol. Bioeng., **22** (1980) 887.
[2] RUKLIŠA, M., VIESTURS, U. and BEKER, M.: Acad. Sci. LSSR, **3** (1976) 90.
[3] RUKLIŠA, M., BEZBORODOV, A. M., VIESTURS, U., SELGA, J. and MEŽINA, I.: Microbiol. Promst. **1** (1973) 36.
[4] RUKLIŠA, M., KAMPE, A., LEITE, M., SALMANE, T. and MARAUSKA, D.: Technology of Microbial Synthesis. Riga, "Zinātne" 1978, 44.
[5] RUKLIŠA, M., JĒKABSONE, M., VIESTURS, U., MEŽINA, I. and SELGA, J.: Prikl. Biokhim. Mikrobiol. **2** (1976) 518.
[6] BEKER, M.: Proc. Conf. on Microbiological and Enzymatic Synthesis of Amino Acids. Puschino, 1980, 5.
[7] BEKER, M., MEŽINA, I., RUKLIŠA, M., VIESTURS, U., SELGA, J., ALEKSANDROVA, M., APSĪTE, A. and SAVENKOV, V. V.: Ist Intern. Symp. on Advances in Microbial Engineering, Praha, 1972, 233.
[8] DUNGAN, J. M., DATTA, P.: J. Biol. Chem. **824** (1973) 8534.
[9] RUKLIŠA, M., ŠVINKA, J., SAKSE, A., VIESTURS, U. and BEKER, M.: Proc. Conf. on Microbiological and Enzymatic Synthesis of Amino Acids. Puschino, 1980, 18.
[10] KARA-MURZA, J. N., IVANOVSKAYA, L. V. and ZHADANOVA, N. U.: Prikl. Biokhim. Mikrobiol. **14** (1978) 345.
[11] SHIIO, I., MIYAJIMA, R.: J. Biochem. **65** (1969) 849.
[12] SANO, K., SHIIO, I.: J. Gen. Appl. Microbiol.: **17** (1971) 248.
[13] SHIIO, I.: J. Ferm. Technol. **53** (1974) 62.
[14] BUKIN, V. M., BEKER, M., KUTSEVA, L. S., BAZDIREVA, M. M., LACARS, A., KĀLVANE, I., SEDVALDS, A., RUKLIŠA, M., BEKERE, V., VIESTURS, U., LIEPIŅŠ, I., VALDMANS, A., VIESTURE, Z., LUZHKOV, M. M. and ARESHKINA, L. Y.: Author. Certificate, USSR No. 255162 (1976).
[15] ŠKAGALE, L., DALKA, R. and UDROVSKY, G.: Mikrobiol. Promst. **3** (1979) 10.
[16] Technical Systems for Microorganism Cultivation (Ed. VIESTURS U.) Riga „Zinātne" 1978, 16.
[17] BLACK, N., WEIGHT, I.: J. Biol. Chem. **213** (1955) 51.
[18] KAGAN, Z. J., MASHATINA, Y. I.: Prikl. Biokhim. Mikrobiol. **5** (1969) 578.
[19] LOWRY, O. H., ROSENBROUGH, N. Y., FARR, A. F., RANDALL, R. J.: J. Biol. Chem. **193** (1951) 104.
[20] MIYAJIMA, R., OTSUKA, J., SHIIO, I.: J. Biochem. **63** (1968) 139.
[21] BEKER, M., RUKLISA, M., ŠVINKA, J., LEITE, M., VIESTURS, U., MARAUSKA, D., SAKSE, A., DUNCE, M. and LIEPIŅŠ, I.: XII Int. Congress of Microbiology. München, 1978, 122.
[22] VIESTURS, U., SELGA, J., SHCHERBAK, N., RUKLIŠA, M. and STŪRMANIS, I.: Microbiological Preparations. Riga „Zinātne", 1976, 176.
[23] KAZUMI ARIKI. USSR Patent No. 298132 (1971).
[24] NAKAYAMA, K., KITADA, J., KINOSHITA, J.: J. Gen. Appl. Microbiol. **17** (1961) 145.

Acta Biotechnologica 2 (1982) 1, 86

BOOK REVIEW

KALUNJANZ, K. A.; JESTAKOW, N. W. and PIWNJAK, I. G.

Die Anwendung von Produkten der mikrobiologischen Synthese in der Tierproduktion

Verlag „Kolos", Moskau, 1980
288 Seiten, 112 Tabellen und Übersichten (in russischer Sprache)

Das Buch befaßt sich mit den Ergebnissen bei der Anwendung von mikrobiell hergestellten Produkten in der modernen industriellen Tierproduktion. Auf die fermentative Herstellung dieser Stoffe wird jeweils kurz eingegangen.
Vorgestellt werden im wesentlichen Ergebnisse und Empfehlungen beim Einsatz von in der Sowjetunion hergestellten Produkten, wobei auch der internationale Stand bzw. internationale Trends berücksichtigt werden. Vorgestellt werden ferner moderne Kontrollmethoden zur Charakterisierung der Produkt- bzw. Futterqualität, Normen über ihren Einsatz in den Rationen landwirtschaftlicher Nutztiere und damit in der industriellen Tierproduktion erzielte Effektivitätskennziffern.
Folgende Produktklassen mikrobieller Herkunft werden betrachtet:

- Futterhefen aus konventionellen und unkonventionellen Rohstoffen
- Aminosäuren
- Antibiotika
- Vitamine
- Enzympräparate

Das Buch wendet sich besonders an die Fachleute auf dem Gebiet der Tierernährung und Tierproduktion, gibt jedoch auch anderen an der Biotechnologie Interessierten die Möglichkeit, sich über die Vielfalt der Anwendungsgebiete mikrobieller Produkte in der modernen Tierproduktion und deren volkswirtschaftlicher Effekte in einem breiten Überblick zu informieren, ohne dabei tiefere theoretische Fachkenntnisse vorauszusetzen.

W. KAURUFF

Acta Biotechnologica 2 (1982) 1, 87—94

Gewinnung und Charakterisierung von Prolidase aus Escherichia coli B.
I. Gewinnung des Enzyms

E. Friese und H. Ruttloff

Akademie der Wissenschaften der DDR
Zentralinstitut für Ernährung
1505 Potsdam-Rehbrücke, Arthur-Scheunert-Allee 114/116

Summary

An enzyme being able to hydrolize the imido linkage at the N-terminal end of proline is isolated from *E. coli B.* This fact corresponds to the specifity of hydrolization of the animal prolidase. Enzyme synthesis within the cells of *E. coli* B is carried out independently from growth. Changed environmental factors may influence the formation rate of enzyme in a restricted way. A relatively high enzyme biosynthesis can be reached by cultivating the strain *E. coli* B at a temperature of 37 °C as well as an initial pH-value of the medium of 7.0 in submerged culture (400—500 rpm). By variation of the medium composition enzyme synthesis does not change considerably, however, biomass yield can be increased about 100%. If mechanical cell desintegration is optimized by means of ultrasonic or vibration homogenisator the cell components may be easily released with a higher proline specific activity as animal prolidase.

Zusammenfassung

Aus *E. coli* B wurde ein Enzym, welches zur hydrolytischen Spaltung der Imido-Bindung am N-terminalen Ende von Prolin befähigt ist, isoliert. Diese Wirkungsspezifität stimmt mit der der tierischen Prolidase überein.
Die Enzymsynthese in den Zellen von *E. coli* B erfolgt wachstumsunabhängig. Die Kultivierungsbedingungen beeinflussen die Enzymbildungsrate entscheidend. Eine relativ hohe Enzymbiosynthese kann bei folgenden Fermentationsbedingungen (submers) erzielt werden: 37 °C; Anfangs-pH-Wert 7,0; Rührgeschwindigkeit 400—500 U/min. Eine veränderte Nährmedienzusammensetzung bewirkt keine entscheidende Veränderung der Enzymsynthese; jedoch kann der Biomasseertrag bis um 100% gesteigert werden. Durch Optimierung der mechanischen Zelldesintegration (Ultraschall oder Vibrohomogenisator) können die Zellbestandteile gewonnen werden; dabei ist eine höhere Prolin-spezifische Aktivität als bei der tierischen Prolidase erreichbar.

Einleitung

Bei der enzymatischen Hydrolyse von Proteinen stellt die Spaltung der Imidobindung zwischen Prolin und einer Aminosäure am N-terminalen Ende vielfach einen limitierenden Schritt dar, da unspezifische Peptidasen hierzu nicht in der Lage sind. Als einziges

Enzym vermag Prolidase diese Bindung zu zerlegen:

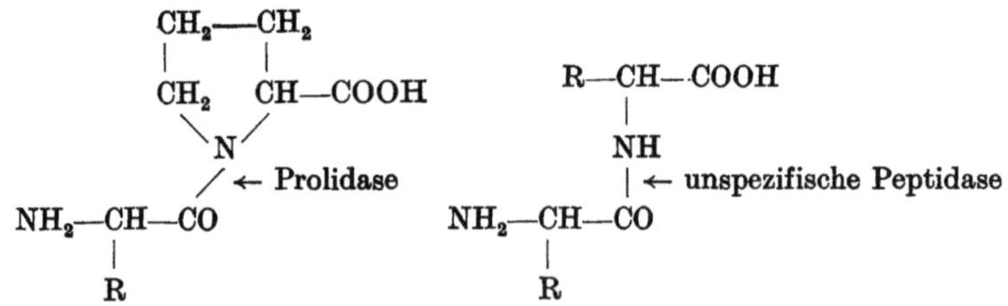

Dieses Problem dürfte insbesondere auch beim Einsatz trägerfixierter Proteasen auftreten, wie sie möglicherweise in Enzymreaktoren zur Totalhydrolyse von Proteasen in Zukunft Verwendung finden.
Prolidase ist in tierischen Geweben vor allem in Verbindung mit dem Metabolismus des Bindegewebes untersucht worden. Von zahlreichen Autoren sind prolinspezifische Enzyme in stoffwechselaktiven Geweben und Organen nachgewiesen und isoliert worden [1, 2, 3, 4, 5, 6, 7].
Für die Gewinnung von Prolidase durch Extraktion aus tierischen Geweben sind erhebliche Mengen an Rohmaterial erforderlich. Die so hergestellten Enzympräparate sind daher kostenaufwendig, und es erscheint geboten, ökonomisch günstigere Rohstoffe für die Prolidasegewinnung zu erschließen. Hierbei ist vorrangig an die Produktion dieses Enzyms durch Mikroorganismen zu denken.
Erste Angaben über die Existenz prolinspezifischer Enzyme in Mikroorganismen gibt STONE [8]. Er weist nach, daß prolinauxotrophe Mutanten von *E. coli* in solchen Nährmedien wachsen, die statt des freien Prolins kurzkettige Prolinpeptide enthalten. Folglich werden kleine Peptide in die Zelle transportiert, und es existieren intrazelluläre Peptidasen, welche die für die Proteinbiosynthese notwendigen Aminosäuren aus den Peptiden freisetzen.
Prolidase wird u. a. in verschiedenen Bakterienstämmen angetroffen. YARON und MLYNAR [9] beschreiben die Herstellung einer intrazellulären „Aminopeptidase P" aus *E. coli* B, die eine ungewöhnliche Substratspezifität besitzt. Sie spaltet N-terminale Aminosäuren ab, wenn diese an Prolin gebunden sind und greift sowohl hoch- als auch niedermolekulare Peptide an; allerdings ist die Aktivität verhältnismäßig niedrig. MCHUGH und MILLER [10] und MOU et al. [11] weisen in *Salmonella typhimuriun* bzw. in *Streptokokken* Enzyme nach, die ebenfalls die Fähigkeit besitzen, die Imidobindung zu zerlegen.
Nachfolgend wird über die Gewinnung von Prolidase aus *E. coli* berichtet.

Material und Methoden

Stammaterial: Verschiedene Species der Stämme *E. coli* SG, *E. coli* B, *E. coli* K, *E. coli* CA, *E. coli* D, *E. coli* WF$^+$, *E. coli* WF$^-$, *E. coli* Hfr aus der Zentralen Stammsammlung des ZIMET Jena.
Nährmedien: Lab-Lemco-Medium: 7 g Pepton, 7 g Hefeextrakt, mit Leitungswasser ad 1000 ml; Einstellung des pH-Wertes auf 7,4 · Medium 19: 7 g Pepton, 7 g Hefeextrakt, 7 g Fleischextrakt; mit Leitungswasser ad 1000 ml; Einstellung des pH-Wertes auf 7,0.
Stammhaltung: Die Stammhaltung von *E. coli* B erfolgt auf Schrägröhrchen mit Lab-Lemco-Agar, die 15 Std. bei 30 °C bebrütet und bei Raumtemperatur bzw. im Kühlschrank aufbewahrt werden.

Vorkultur: Ein Schrägrohr wird mit 5 ml steriler physiologischer Kochsalzlösung abgeschwemmt, diese in 100 ml Schüttelmedium (Medium 19, 500 ml-Rundkolben) übertragen und 15 Std. bei 30 °C auf einem Rundschwingtisch (300 Upm) inkubiert.
Arbeitskultur: Beispiel 1. 20 ml der Vorkultur werden in 500 ml Medium 19 (1 l-Erlenmeyerkolben) überimpft. Anschließend wird 4 bis 5 Std. lang bei 37 °C geschüttelt (Rundschwingtisch 300 Upm). Die gebildete Zellmasse wird durch Zentrifugieren von der Kulturflüssigkeit abgetrennt.
Beispiel 2. Aus der Vorkultur wird in 30 l-Fermentoren überimpft, die 20 l Medium 19 enthalten (2% Inokulum). Die Kultivierung erfolgt bei 37 °C und einer Rührgeschwindigkeit von 350 Upm, die Luftzufuhr beträgt 10 l Luft je min auf 20 l Flüssigkeit. Nach 4 bis 5 Std. wird die Zellmasse mittels Separator von der Kulturflüssigkeit abgetrennt.
Wachstumsmessung: Die Zunahme der Biomasse wird durch Trübungsmessung der Kulturflüssigkeit bestimmt. Hierzu werden 3 ml Kulturmedium mit 1 ml Formaldehyd versetzt umd die Extinktion bei 700 nm (Spekol, VEB Carl Zeiss Jena) gemessen.
Zellaufschluß: Die Zerstörung der Zellen erfolgt mechanisch mittels Ultraschall (20 KHz) bzw. Vibrationshomogenisator (Schütteln mit Glaskugeln, 20 KHz). Nähere Angaben werden im Ergebnisteil dargestellt.
Messung der Enzymaktivität: Sie erfolgt durch Bestimmung des freigesetzten Prolins mit Ninhydrinreagenz [12, 13].
Bestimmung des Prolingehaltes: Es wird die Methode nach LOWRY et al. [14] angewandt.
Bestimmung des Stickstoffgehaltes: Der Stickstoffgehalt der flüssigen Nährmedien wird nach einer Methoden-Kombination KJELDAHL-Naßaufschluß/BERTHELOT-Reaktion mit Hilfe einer teilautomatisierten Analysenlinie bestimmt [15].

Ergebnisse und Diskussion

Zur Auswahl eines geeigneten Enzymbildners werden 70 *E. coli*-Stämme (vgl. Material und Methoden) hinsichtlich Prolidase-Bildung unter konstanten Bedingungen getestet (Lab-Lemco-Medium, Kultivierung 6 Std. bei 30 °C, Ultraschallaufschluß). Von den getesteten Stämmen weisen 22 prolinspezifische Aktivitäten auf. Die höchste Prolidaseaktivität wird in einem *E. coli* B nachgewiesen. Dieser Stamm wird deshalb für die weiteren Untersuchungen zur Optimierung herangezogen. Der Entwicklungsstand der Stammkultur (Schrägrohr: Lab-Lemco; Bebrütungszeit: 6 bis 20 Std.; Stammlagerung: 1 bis 4 Wochen) ist ohne Einfluß auf die Biosyntheseleistung in der Arbeitskultur. Günstig erweist sich indessen die Führung einer Vorkultur unter Verwendung eines Mediums gemäß Arbeitskultur (Kultivierung 6 bis 15 Std. bei 30 °C), wodurch die exponentielle Wachstumsphase in der Arbeitskultur verkürzt wird.

1. Optimierung des Nährmediums der Arbeitskultur

Die Nährmedienoptimierung erfolgt unter Einsatz eines Komplexmediums. Nach den Ergebnissen einer empirisch durchgeführten Voruntersuchung im Hinblick auf die Zusammensetzung und Konzentration der Nährstoffe werden folgende Komponenten für die Optimierung in einem statistischen 2^5-Versuchsplan nach BOX-WILSON ausgewählt:

	Konzentration im Medium [%]
1. Hefeextrakt	0,7
2. Fleischextrakt	0,7
3. Pepton	0,7
4. Maisquellwasser	0,7
5. Glucose	1,0

Bei Vorgabe der verschiedenen Medien sind große Unterschiede in der Enzymsynthese nicht zu erkennen. Jedoch läßt sich die Biomasseproduktion in den Medien 1, 3, 4, 7, 17 und 19 gegenüber dem bisher verwendeten Lab-Lemco-Medium (4,0 g Feuchtmasse) um 100% steigern. Medium 19 (enthält Fleischextrakt, Hefeextrakt, Pepton) wird für die weiteren Untersuchungen ausgewählt (Abb. 1).

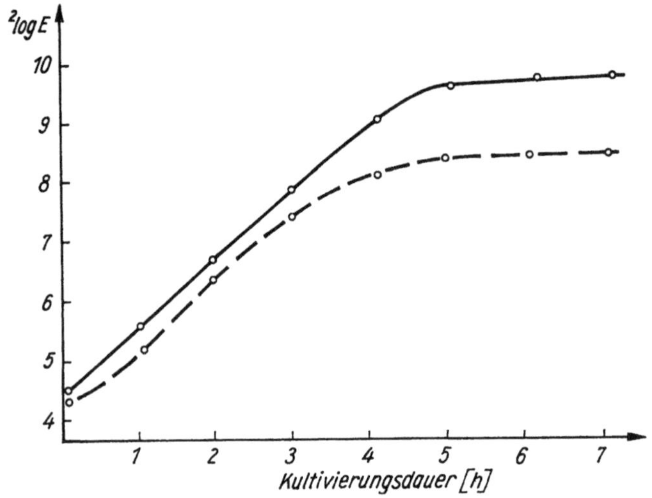

Abb. 1. Wachstum in Lab-Lemco-Medium sowie in Medium 19 (– – – Lab-Lemco, —— Medium 19).

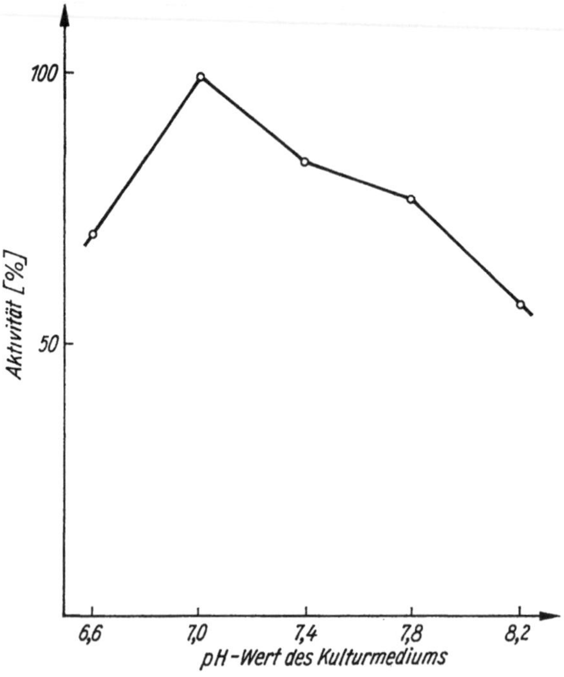

Abb. 2. Abhängigkeit der Prolidasebildung vom pH-Wert des Mediums 19 (37 °C, 6 Std., 300 Upm, Ultraschallaufschluß)

2. Einfluß des Ausgangs-pH-Wertes auf die Enzymbildung

Untersuchungen zur Abhängigkeit der Enzymproduktion vom Ausgangs-pH-Wert des Mediums ergeben die in Abbildung 2 dargestellten Ergebnisse. Danach liegt der optimale pH-Wert beim Neutralpunkt.

3. Ermittlung der optimalen Kultivierungstemperatur

In Versuchen zur Optimierung der Kultivierungstemperatur werden Ergebnisse gemäß Abbildung 3 erhalten. Die günstigste Temperatur ist danach 37 °C; Abweichungen hiervon sind stets mit einem Aktivitätsabfall verbunden.

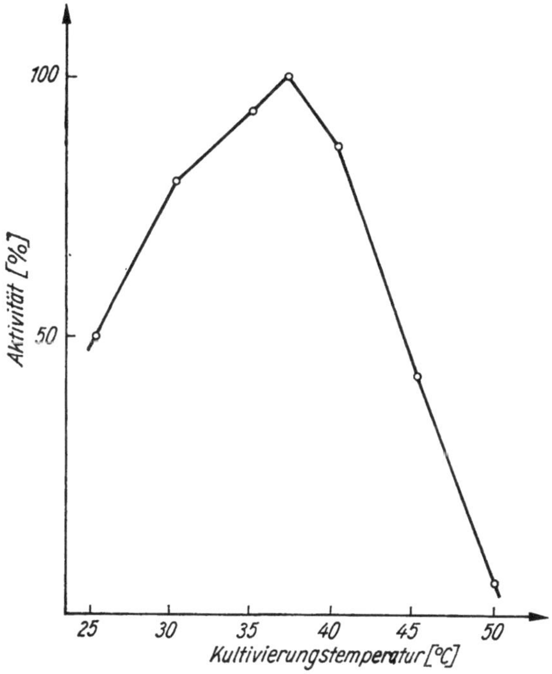

Abb. 3. Abhängigkeit der Prolidasebildung von der Kultivierungstemperatur des Mediums 19 (6 Std., 300 Upm, Ultraschallaufschluß)

4. Einfluß der Kultivierungsdauer

Unabhängig von der Wachstumsphase werden in den Zellen stets gleich hohe Prolylpeptid-spezifische Enzymaktivitäten nachgewiesen, d. h. die Kultivierungsdauer ist auf die intrazelluläre Prolidasekonzentration ohne Einfluß (Abb. 4).

5. Belüftung und Rührgeschwindigkeit

Die Versuchsansätze zur Optimierung von Belüftung und Rührgeschwindigkeit werden in 30 l-Fermentoren durchgeführt. Beide Parameter haben in den getesteten Bereichen keinen Einfluß auf die intrazelluläre Enzymbildung, sie beeinflussen jedoch die Wachstumsgeschwindigkeit. Die günstigsten Bedingungen für das Wachstum liegen bei Einhaltung folgender Parameter vor: Rührung im Bereich zwischen 450 und 500 Upm; Belüftung: 10 l/min; Fermentorfüllung: 20 l; Temperatur: 37 °C; Anfangs-pH-Wert des Mediums 19: pH 7,0; Entschäumerzusatz: 0,1 bis 0,2% Rapsöl.

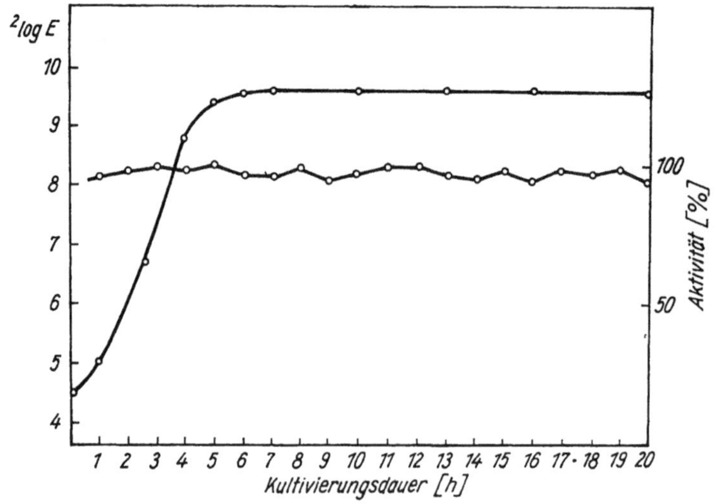

Abb. 4. Prolidasebildung im Verlauf des Wachstums von *E. coli* B (Medium 19; 37 °C; 20 Std.; 300 Upm)

6. Zellaufschluß

Die Zerstörung der *E. coli*-Zellen wird mechanisch mittels Ultraschall bzw. Vibrationshomogenisator durchgeführt. Zur Ermittlung der optimalen Bedingungen für die Ultraschallbehandlung werden die Beschallungsdauer, das zu behandelnde Volumen sowie die Dichte des Volumens variiert. Die Beschallfrequenz wird konstant auf 20 KHz gehalten (maximale Leistung des Gerätes nach KERRY HG 100). Folgende Parameter werden als optimal gefunden: Volumen = 10 ml; Beschallungsdauer = 5 min; Dichte des Volumens = 1 g Feuchtgewicht in 10 ml Puffer.

Als Optimalbedingungen für die Zerstörung der Zellen mittels Glaskugel-Vibrationshomogenisator werden ermittelt: Durchmesser der Ballotini = 0,1 mm; Menge der Ballotini = 8 g; Biomasse = 2 g feuchte Zellen; Behandlungsdauer: 2 min. Die Frequenz des Vibrationshomogenisators (VHG 1) beträgt 20 KHz. Einen Vergleich der nach den genannten optimierten Zellaufschlußmethoden erzielten Ausbeuten zeigt Abbildung 5.

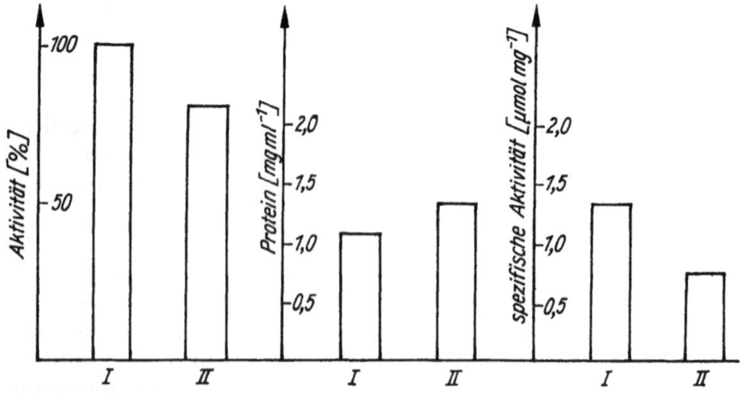

Abb. 5. Vergleich der Ausbeuten nach Zellaufschluß mittels Vibrationshomogenisator bzw. Ultraschall (I = Vibrationshomogenisator, II = Ultraschall)

7. Fermentationsverlauf

Zur Charakterisierung der Fermentation werden Wachstum, pH-Verlauf, Veränderung des Gesamtstickstoff- und Proteingehaltes im Kulturmedium über einen Zeitraum von 20 Std. verfolgt (Abb. 6).
Der stärkste Abbau der Proteine bzw. des Gesamtstickstoffs erfolgt im Verlauf des logarithmischen Wachstums. Der pH-Wert des Mediums bleibt im großen und ganzen konstant. Die Enzymkonzentration in der Zelle des Stammes *E. coli* B ist unabhängig von der Wachstumsphase der Kultur; sowohl in der logarithmischen als auch in der stationären Phase werden in der Biomasse etwa gleich hohe Aktivitäten an Prolylpeptid-spezifischem Enzym nachgewiesen.

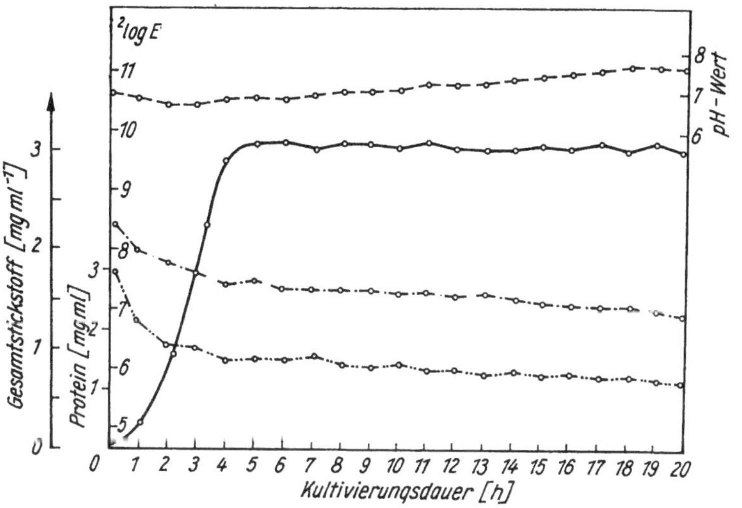

Abb. 6. Wachstum der Kultur, pH-Wert, Gehalt des Mediums an Gesamtstickstoff und Protein bei der Fermentation im 30-l-Gefäß (Rührung: 450 Upm, Belüftung: 10 l/min, Medium 19; —— Wachstum, - - - pH-Verlauf, -·-·- Protein, -··-··- Gesamtstickstoff)

Schlußbemerkungen

Eine Versuchsdurchführung im 300 l-Fermentor mit einer Füllmenge von 200 l Nährmedium liefert 2 kg Biomasse-Feuchtsubstanz. Nach Zellaufschluß mittels Vibrationshomogenisator beträgt die Ausbeute etwa 20 g lyophilisiertes Rohenzym mit einer spezifischen Aktivität von 1 μmol/min mg gegenüber dem Substrat Glycyl-Prolin. Somit spaltet die *E. coli*-Prolidase (Rohenzym) das Substrat Gly-Pro mit etwa der 30fachen Geschwindigkeit gegenüber der Spaltungsrate der von Yaron und Mlynar [9] ebenfalls aus *E. coli* B isolierten Aminopeptidase P (spezifische Aktivität: 0,064 μmol/min mg).

Eingegangen: 26. 1. 1981

Literatur

[1] Adams, E., Smith, E. L.: J. biol. Chem. **198** (1952) 671.
[2] Barth, A., Hermann, G.: Biochem. Physiol. Pflanzen **166** (1974) 23.
[3] Barth, A., Schulze, H., Neubert, K.: Acta biol. med. german. **32** (1974) 157.
[4] Davis, N. G., Smith, E. L.: J. biol. Chem. **224** (1957) 261.
[5] Dehm, P., Nordwig, A.: Eur. J. Biochem. **17** (1970) 372.
[6] Sjöström, H.: Acta Chem. Scand., Ser. B **28** (1974) 802.
[7] Smith, E. L.: Advances in Enzymol. **12** (1951) 191.
[8] Stone, D.: J. biol. Chem. **202** (1953) 821.
[9] Yaron, A., Mlynar, D.: Biochem. biophys. Res. Commun. **32** (1968) 658.
[10] McHugh, G. L., Miller, C. G.: J. Bacteriol. **120** (1974) 364.
[11] Mou, L., Sullivan, J. J., Jago, G. R.: J. Dairy Res. **42** (1975) 147.
[12] Troll, W., Lindsley, J.: J. biol. Chem. **215** (1955) 655.
[13] Kirschke, H.: Dissertationsschrift der Martin-Luther-Universität Halle (1966).
[14] Lowry, O. H., Rosebrough, N. J., Farr, A. L., Randall, R. J.: J. biol. Chem. **193** (1951) 265.
[15] Lange, R., Friebe, R., Linow, F.: Nahrung **23** (1979) 549.

Acta Biotechnologica 2 (1982) 1, 95–102

Erzeugung eines zellwandlytischen Enzymkomplexes aus Arthrobacter GJM-I zur Herstellung von Protoplasten aus Candida spec. H

O. Hässler, H.-G. Wisniewski, B. Röber und G. Reuter

Wissenschaftsbereich Mikrobielle Biochemie der Sektion Biologie,
Friedrich-Schiller-Universität Jena,
DDR-6900 Jena, Wöllnitzer Str. 7

Summary

Growing conditions have been found out for the bacterium *Arthrobacter* GJM-I to produce a lytic enzyme system, which converts cells of the yeast *Candida spec.* H to protoplasts quickly and in a good yield. Estimating the activities of α-mannanase and β-glucanase we found out the optimal culture time to gain the lytic enzyme system from the culture filtrate. It was shown that radioactive labeling of the yeast cells makes it possible to estimate quantitatively the conversion to protoplasts and the simultaneous lysis. The obtained lytic enzyme system can substitute the snail enzyme system which was used for cell conversion of *Candida spec.* H to protoplasts till now.

Zusammenfassung

Für den Bakterienstamm *Arthrobacter* GJM-I wurden Wachstumsbedingungen ermittelt, um ein Enzympräparat zu gewinnen, das aus Zellen der Hefe *Candida spec.* H rasch und in guter Ausbeute Protoplasten erzeugt. Durch Bestimmung der α-Mannanase- und β-Glucanase-Aktivität wurde die zur Enzymgewinnung aus dem Kulturfiltrat geeignete Kulturzeit ermittelt. Durch radioaktive Markierung der Hefe wurde neben der Protoplastenbildung die gleichzeitig auftretende Lyse quantitativ verfolgt. Das erhaltene lytische Enzymsystem kann das bisher verwendete Schneckenenzym bei der Protoplastengewinnung von *Candida spec.* H ersetzen.

Einleitung

Zum Abbau von Zellwänden lebender Hefen wird in der Regel das aus dem Verdauungssaft von *Helix pomatia* gewonnene Schneckenenzym verwendet [1]. Da dieses nur beschränkt zugänglich und in seiner Qualität nicht einheitlich ist, wurde wiederholt versucht, lytische Enzyme mikrobiellen Ursprungs für diesen Zweck einzusetzen [2, 3, 4, 5, 6,].

Eine besonders hohe lytische Aktivität gegen lebende Hefezellen weisen Stämme der Gattung *Arthrobacter* auf [7, 8, 9, 10, 11]. Der aus Abwässern isolierte Bakterienstamm *Arthrobacter* GJM-I ist beim Wachstum auf Mannan von *Saccharomyces cerevisiae* ein guter Produzent von extrazellulären Enzymen, hauptsächlich von α-Mannanase. Aus diesem Grund wurde er auch bei Strukturuntersuchungen von Mannan eingesetzt [12].

Andere setzten das komplette lytische Enzymsystem von *Arthrobacter* GJM-I, bestehend aus α-Mannanase, β-1,3-Glucanase und Protease, zur Protoplastenpräparation und Lyse bei verschiedenen Hefen ein [13, 14]. Unter diesen Bedingungen wurde bei *Saccharomyces cerevisiae* eine hohe Protoplastenausbeute erzielt. Zellen von *Candida*-Arten dagegen ließen sich mit *Arthrobacter*-Enzym nur in geringen Ausbeuten zu Protoplasten konvertieren.

Als Voraussetzung für Untersuchungen zur Biosynthese der Hefezellwand sollten Protoplasten hergestellt werden aus *Candida spec.* H, einer von uns untersuchten Hefe, die auf *n*-Alkanen als einziger C- und Energiequelle zu wachsen vermag. Durch Auswahl entsprechender Kulturbedingungen haben wir aus *Arthrobacter* GJM-I einen Enzymkomplex erhalten, der erfolgreich zur Protoplastengewinnung aus Candida *spec.* H eingesetzt werden kann [15]. Quantitative Aussagen zur Protoplastenausbeute, zur Einschätzung der Lyse während der Protoplastenherstellung sowie Vergleiche zwischen *Arthrobacter*- und Schneckenenzym wurden durch ^{14}C-Leucin-Markierung während der Hefe-Anzucht möglich.

Material und Methoden

Mikroorganismen:

Arthrobacter GJM-I: Institut für Chemie der Slowakischen Akademie der Wissenschaften, Bratislava, CSSR.
Candida spec. H: Institut für Technische Chemie der Akademie der Wissenschaften der DDR, Leipzig.[1])
Saccharomyces cerevisiae: handelsübliche Bäckerhefe.

Schneckenenzym:

Der Verdauungssaft von *Helix pomatia* wurde aus den Mitteldarmdrüsen ausgepreßt, vorfiltriert, sterilfiltriert und lyophilisiert.

Isolation der Hefezellwand:

Die Hefe wurde in wenig Aqua dest. suspendiert und in einer DYNO-MILL (W. Bachofen Maschinenfabrik, Basel, Schweiz) mit Ballotini im kontinuierlichen Verfahren zerstört. Der Zerstörungsgrad wurde durch Methylenblaufärbung kontrolliert. Die Zellwände wurden bei 5000 g (5 °C, 15 min) abzentrifugiert, dreimal mit Aqua dest. gewaschen und lyophilisiert.

Arthrobacter-Kulturen:

Stammhaltung: 10 g Hefeextrakt, 10 g Pepton, 35 g Agar und 1 g Mannan mit Aqua dest. auf 1000 ml.
Flüssigmedium: 7,5 g K_2HPO_4, 2,32 g KH_2PO_4, 3,75 g $(NH_4)_2SO_4$, 1,5 g Hefeextrakt, 20 g Kohlenhydrat (vgl. Versuchsbedingungen), 10 mg $FeSO_4 \cdot 7H_2O$, 300 mg $MgSO_4 \times 7H_2O$, 75 mg $CaCl_2 \cdot 6H_2O$, mit Aqua dest. auf 1000 ml.
Der Bakterienrasen zweier Kulturröhrchen wurde in 200 ml Flüssigmedium eingeschwemmt und dieses 48 h bei 28 °C geschüttelt. 10 ml der Vorkultur wurden in 200 ml Hauptkultur überführt und diese mindestens 2 Tage bei 28 °C geschüttelt (Rundschüttler K12, VEB Kyffhäuserhütte, Artern).

1 (Nach persönlicher Mitteilung von Dr. Sattler, ItC Leipzig, steht der zur Verfügung gestellte Stamm einem von ItC inzwischen als *Lodderomyces elongisporus* identifizierten Stamm sehr nahe.)

Isolation des lytischen Enzymkomplexes:

Die *Arthrobacter*-Kulturen wurden bei 5000 g (30 min, 5 °C) zentrifugiert und die Überstände im Vakuum bei 5 °C auf 1/10 eingeengt. Das Konzentrat wurde sterilfiltriert und anschließend 2 Tage gegen Aqua dest. dialysiert (SERVA-VISKING-Dialyseschlauch, Ausschlußgröße 10000—20000 Dalton). Das nach der Lyophilisierung erhaltene Rohenzym ist ein weißes Pulver, das zu etwa 30—50% aus Protein besteht.

Bestimmung der Mannanase- und Glucanaseaktivität:

Mannan und Glucan, gewonnen aus *Candida spec.* H und *S. cerevisiae,* wurden enzymatisch gespalten und die reduzierenden Einheiten anschließend nach dem SOMOGYI-NELSON-Test bestimmt.
Die Ansätze von 500 μl enthalten: 250 μl 0,1 M Phosphatpuffer nach SÖRENSEN mit einem pH-Wert von 7,6 zum Nachweis der Mannanase- bzw. 5,0 zum Nachweis der Glucanaseaktivität, 200 μl einer 1 mg/ml enthaltenden Mannanlösung bzw. Glucansuspension und 50 μl *Arthrobacter*-Enzymlösung. Die Ansätze wurden 1 h bei 30 °C inkubiert und die enzymatische Hydrolyse durch Zugabe von 1 ml der alkalischen Kupferreagenslösung gestoppt. Als Blindansätze wurden Proben mit inaktiviertem Enzym zur Kontrolle mitgeführt. In den Ansätzen erfolgte anschließend eine Bestimmung der freigesetzten reduzierenden Zucker [16]. Die Proteinbestimmung erfolgte in Anlehnung an [17]. Eine Enzymeinheit (U) ist dabei die Enzymaktivität, die bei 30 °C in einer Minute 1 μmol monomeres Produktäquivalent freisetzt.

Anzucht von Candida spec. H zur Zellwandgewinnung:

20 g Glucose, 5 g Hefeextrakt, 3 g $(NH_4)_2SO_4$, 1 g KH_2PO_4, 0,16 g K_2HPO_4, 0,7 g $MgSO_4 \times 7H_2O$, 0,5 g NaCl, 0,4 g $Ca(NO_3)_2$ mit Aqua dest. auf 1000 ml, pH — 5,4, Kulturdauer im Fermentor 24 h bei 28 °C.

Anzucht von Candida spec. H *zur Protoplastengewinnung*:

Die Hefe wurde bei 28°C in Schüttelkulturen auf einem Flüssigmedium kultiviert: 0,5 g KH_2PO_4, 0,5 g $MgSO_4 \cdot 7H_2O$, 0,3 g $CaCl_2 \cdot 2H_2O$, 3,0 g $(NH_4)_2SO_4$, 20 g Glucose, 2 g L-Asparagin (oder 0,45 g Harnstoff), 0,1 mg KJ, 60 mg $MnSO_4 \cdot 4H_2O$, 300 mg $ZnSO_4 \cdot 7H_2O$, 40 mg $CuSO_4 \cdot 5H_2O$, 25 mg $Na_2MoO_4 \cdot 2H_2O$ und 240 mg $FeCl_3 \cdot 6H_2O$ mit Aqua dest. auf 1000 ml [18].

Präparation der Protoplasten:

Die abzentrifugierten Hefezellen wurden mit 1prozentiger Kaliumchloridlösung gewaschen und anschließend 30 min bei 28 °C in 40 ml Vorinkubationsmedium (700 mg L-Cystein, 744 mg EDTA, 61,6 mg 2,4-Dithioerythrit mit Aqua dest. auf 40 ml, mit 1 N NaOH auf pH 6 eingestellt) geschüttelt, anschließend zweimal in 0,1 M Phosphatpuffer (pH 6, mit 0,6 M Mannit osmotisch stabilisiert) gewaschen und bis zu 200 mg/ml Hefefeuchtmasse im gleichen Puffer suspendiert. Nach Zusatz von *Arthrobacter*-Enzymkomplex (bis 20 mg/ml) wurde 20 min bei 28 °C auf einem Reziprokschüttler inkubiert.

Bestimmung von Protoplastenausbeute und Lyse:

In Vorversuchen zur Bestimmung der Inkorporationskinetik von ^{14}C(U)-L-Leucin an *Candida spec.* H nach Zugabe zum harnstoffhaltigen Kulturmedium wurde festgestellt, daß die ^{14}C-Gesamtinkorporation (gemessen nach Protosolaufschluß der Zellen) nach 180 min einen Maximalwert von 52% der eingesetzten Radioaktivität erreicht. Deshalb

wurden den Kulturen in der logarithmischen Phase des Wachstums, 3 h vor dem Abernten, 5 μCi ^{14}C(U)-L-Leucin (185 mCi · mmol^{-1}) zugesetzt.
Nach der Vorinkubation wurden aus der gewaschenen radioaktiven Hefe zwei volumengleiche Hefesuspensionen mit 50—200 mg/ml Feuchtmasse im Präparationspuffer hergestellt, die als Ansatz (mit Enzymkomplex, bis zu 20 mg/ml) bzw. als Kontrolle Verwendung fanden. Aus den bei 28 °C geschüttelten Suspensionen wurden in bestimmten zeitlichen Abständen je 200 μl Proben in je 1 ml osmotisch stabilisierten Präparationspuffer, 1 ml nicht stabilisierten Präparationspuffer mit 10% Triton X-100 sowie zu 2 ml Protosol pipettiert. Die Protosolproben dienten zur Bestimmung der ^{14}C-Gesamtinkorporation. Sie wurden 20 h bei 55 °C inkubiert und anschließend in 10 ml Dioxanszintillator in Probegläser zur Szintillationsmessung überführt. Die übrigen Proben wurden sofort 10 min bei 5600 g zentrifugiert.
Je 500 μl der Überstände dieser Proben wurden in 10 ml Dioxanszintillator gegeben. Die Radioaktivitätsbestimmung erfolgte in einem LKB Wallac 81000 liquid scintillation counter. Im einzelnen gilt für die Freisetzung der Radioaktivität (A) in der vorgegebenen Zeiteinheit:

A_1 = im Überstand der Kontrolle (ohne Enzymzusatz), osmotisch stabilisiertes Medium (Freisetzung durch Membranpermeabilität, Turnover etc.)

A_2 = im Überstand vom Ansatz (mit Enzym), osmotisch stabilisiertes Medium (Freisetzung durch Permeabilität und Lyse von Zellen einschließlich Freisetzung entspr. A_1)

A_3 = im Überstand vom Ansatz (mit Enzym), unstabilisiertes Medium $\triangleq$ Gesamtfreisetzung (Freisetzung durch Platzen der bei A_2 entstandenen Protoplasten einschließlich der Freisetzung entspr. A_1)

$\bar{A}_0$ = nach Protosolbehandlung der Kontrolle und der Ansätze ohne Zentrifugation $\triangleq$ durchschnittliche Gesamt-Inkorporation als Bezugsgröße für die Anteile freigesetzter Radioaktivität zur quantitativen Einschätzung der Aktivität des Enzympräparates im Hinblick auf die Gewinnung von Protoplasten.

$$A_{r3} = \frac{A_3}{\bar{A}_0} \cdot 100\%$$ (relative Gesamtfreisetzung radioaktiven Zellmaterials während der Enzymeinwirkung)

$$A_{r2} = \frac{A_2 - A_1}{\bar{A}_0} \cdot 100\%$$ (relative Freisetzung radioaktiven Zellmaterials durch Permeabilisierung und Lyse von Zellen während der Enzymeinwirkung)

$$A_{r1} = \frac{A_3 - A_2}{\bar{A}_0} \cdot 100\%$$ (relative Freisetzung aus Protoplasten nach Enzymeinwirkung)

Ergebnisse und Diskussion

Ermittlung geeigneter Bedingungen für die Enzymbildung:

Mit Kulturfiltrat von *Arthrobacter* wurden unmittelbar Zellen von *Saccharomyces cerevisiae* lysiert [13]. Die ungünstigen Ergebnisse mit *Candida*-Arten schienen sich zunächst auch bei dem von uns bearbeiteten Stamm zu bestätigen. Unsere Versuche, Zellen von *Candida spec.* H nach Vorinkubation in EDTA-Dithioerythrit-Medium mittels Kulturfiltrat von *Arthrobacter* (Kulturdauer 70 h, 1 g *S. cerevisiae*-Zellwand pro Liter) in Protoplasten umzuwandeln, brachten keinerlei Erfolg. Eine gute Protoplastenausbeute wurde erst durch eine 100fache Steigerung der eingesetzten Enzymaktivitäten erreicht.
Um zu höheren Enzymausbeuten zu kommen, wurden Versuche durchgeführt mit qualitativer und quantitativer Veränderung der C-Quelle im Kulturmedium unter Berücksichtigung der Enzymaktivitäten in Abhängigkeit von der Kulturdauer.

Die Zellwand von *S. cerevisiae* wurde in steigenden Konzentrationen im Kulturmedium suspendiert und die Änderung der lytischen Aktivität während der Kultivierung verfolgt.
Erhielten wir beim Einsatz von 1 g Zellwand/Liter Kulturlösung und einer Kulturdauer von 70 Stunden eine Mannanseaktivität von nur 3,4 mU/ml Kulturfiltrat, so steigt bei einer Kultur mit 20 g Zellwand/Liter und einer Kulturdauer von 172 Stunden die Mannanaseaktivität des Kulturfiltrates auf 135 mU/ml an (Abb. 1). In einer Vergleichskultur wurden sogar Werte um 200 mU/ml erzielt.

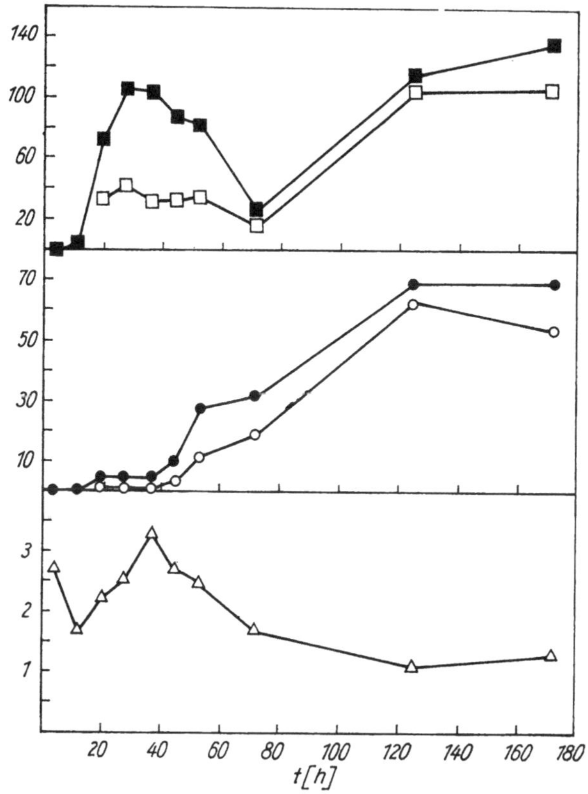

Abb. 1. Proteingehalt [mg/ml] (△—△) und Enzymaktivität (Freisetzung reduzierender Zucker aus Mannan resp. Glucan von *S. cerevisiae*) im Kulturfiltrat von *Arthrobacter* GJM-I während der Kultivierung auf 20 g *S. cerevisiae*-Zellwand pro Liter.
Glucanase [mU/ml] (●—●), [mU/mg] (○—○),
Mannanase [mU/ml] (■—■), [mU/mg] [□—□).

In einer Kultur mit 30 g Zellwand/Liter erreichten wir dagegen nach einer Kulturdauer von 172 Stunden nur eine Mannanaseaktivität von 105 mU/ml mit einer deutlich erniedrigten spezifischen Aktivität. Die Zellwand wurde in dieser Kultur nur unvollständig abgebaut.
Die höchsten Werte für die Glucanaseaktivität wurden gleichfalls in der Kultur mit 20 g Zellwand/Liter ermittelt. Nach einer Kultivierungszeit von 172 Stunden betrug sie 70 mU/ml (Abb. 1).
In nachfolgenden Versuchen wurden die polymeren C-Quellen unterschiedlicher Herkunft hinsichtlich ihrer Beeinflussung der *Arthrobacter*-Enzymaktivität verglichen. Wie

bereits in Abb. 1 (*S. cerevisiae*-Zellwand als C-Quelle) zum Ausdruck kam, durchläuft die Mannanase-Aktivität während der ersten 30 Kulturstunden einen Maximalwert, der mit dem maximalen Proteingehalt des Kulturfiltrats korreliert. Dies gilt auch für Zellwand aus *Candida. spec.* H sowie für die autoklavierte gepreßte Bäckerhefe (Abb. 2). Nach einem Absinken der Enzymaktivität ist bei den Kulturen mit isolierten Zellwandpräparaten ein erneuter Anstieg der Mannanase-Aktivität zu verzeichnen, der nach 170-stündiger Kulturzeit den ersten Maximalwert sogar übertrifft.
Die in Schüttelkulturen zur Enzymgewinnung ermittelten Bedingungen ließen sich inzwischen auf Fermentoren übertragen.

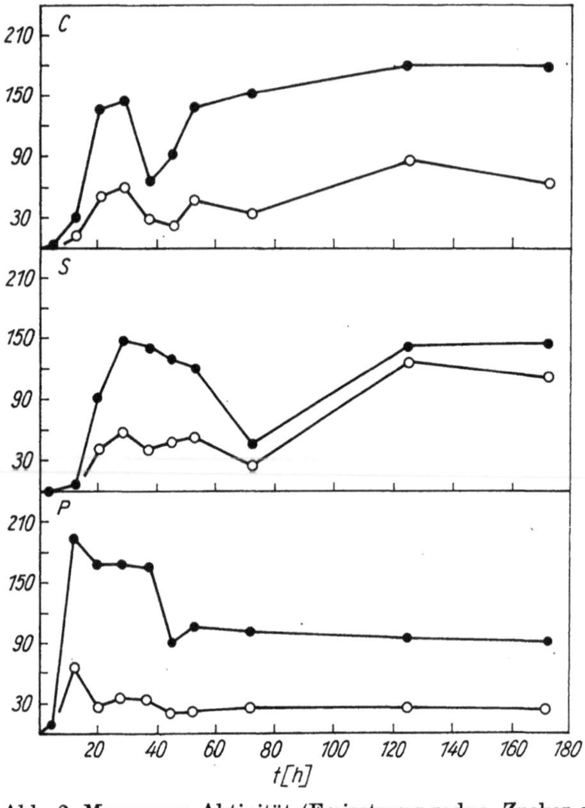

Abb. 2. Mannanase-Aktivität (Freisetzung reduz. Zucker aus Mannan von *Candida spec.* H) im Kulturfiltrat von *Arthrobacter* GJM-I während der Kultivierung auf unterschiedlichen C-Quellen:

C = *Candida spec.* H, lyophilisierte Zellwand, 20 g/l Nährlösung. *S* = *Saccharomyces cerevisiae*, lyophilisierte Zellwand, 20 g/l Nährlösung. P = Preßhefe, autoklavierte gepreßte Bäckerhefe, Feuchtmasse 100 g/l Nährlösung. Enzymaktivität im Kulturfiltrat [mU/ml] (●—●), spezifische Enzymaktivität [mU/mg] (○—○).

Quantitativ vergleichende Gewinnung von Protoplasten mit zellwandlytischen Enzymen:

Zur Präparation von Protoplasten aus *Candida spec.* H eignen sich am besten Zellen einer 6-h-Kultur aus der mittleren log-Phase [18]. Nach Behandlung im Vorinkubationsmedium wurden nach 20 min mit *Arthrobacter*-Enzym die Protoplasten in etwa 95prozentiger Ausbeute erhalten. Die Konzentration der eingesetzten Hefe (Feuchtmasse) betrug dabei 60 mg/ml; die des eingesetzten Enzymproteins 2,8 mg/ml mit einer Aktivi-

tät gegenüber *Candida*-Mannan von 900 mU/ml und eine Aktivität gegenüber *Candida*-Glucan von 60 mU/ml.
Wurden von *Candida spec.* H dagegen Zellen aus der späteren log-Phase zur Protoplastenpräparation herangezogen, so sank die Protoplastenausbeute auf Werte um 60 Prozent ab und es kam zu einer verstärkten Lyse der Zellen. Dabei wurde die Abnahme der ^{14}C-Freisetzung aus Protoplasten durch einen Anstieg der ^{14}C-Freisetzung durch Lyse quanti-

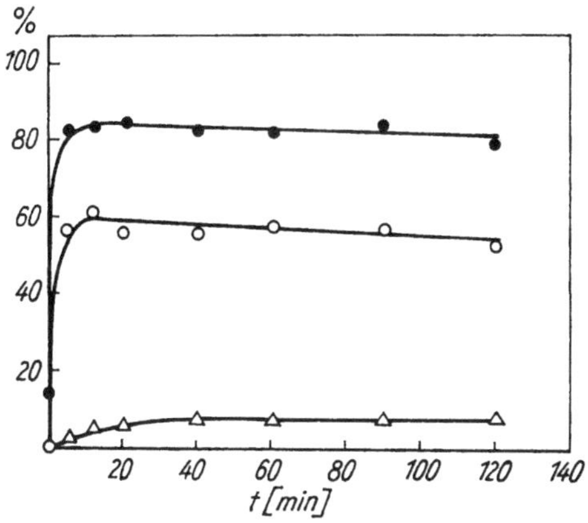

Abb. 3. ^{14}C-Freisetzung (in % der Gesamtinkorporation) bei unterschiedlicher Inkubationszeit aus Protoplasten in Gegenwart von Triton (●—●), aus Protoplasten ohne Triton (○—○), durch Lyse während der Inkubation (△—△). Hefe (berechnet als Trockenmasse) im Ansatz: 9,7 mg/ml. *Arthrobacter*-Enzymkomplex (lyophilisiertes Präparat) im Ansatz: 40 mg/ml.

tativ kompensiert. Die ^{14}C-Freisetzung aus Protoplasten und durch Lyse variiert im Bereich der eingesetzten Hefetrockenmasse (9,7 mg/ml bzw. 32 mg/ml) und der Enzymkonzentration (40 mg/ml bzw. 10 mg/ml) nur unbedeutend. Es ändert sich lediglich die Zeit bis zum vollständigen Umsatz (Abb. 3 u. 4).
Vergleicht man *Arthrobacter*-Enzym und Schnecken-Enzym hinsichtlich seiner spezifischen Mannanaseaktivität und seiner Eignung zur Protoplastenpräparation aus Zellen von *Candida spec.* H, so kann festgestellt werden, daß *Arthrobacter*-Enzym eine 100fach

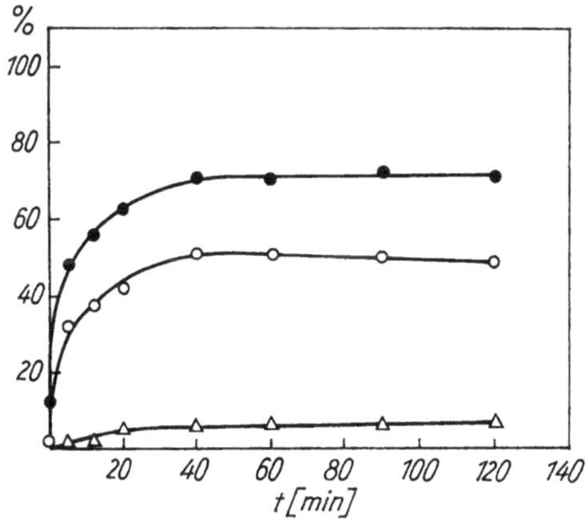

Abb. 4. ^{14}C-Freisetzung (vgl. Abb. 3). Hefe im Ansatz: 32,0 mg/ml, *Arthrobacter*-Enzymkomplex im Ansatz: 10 mg/ml.

höhere spezifische Mannanaseaktivität besitzt. Dies mag in kausalem Zusammenhang damit stehen, daß von uns mit dem *Arthrobacter*-Enzym in kürzerer Inkubationszeit höhere Ausbeuten an Protoplasten erzielt werden konnten als mit Schneckenenzym. Dennoch ist die Beteiligung der Mannanase an der Lyse der Zellwand sicher nur im Zusammenhang mit der Wirkung der anderen im Komplex nachzuweisenden Enzyme [13, 14] zu sehen.

Eingegangen: 29. 12. 1980

Literatur

[1] Eddy, A. A. and Williamson, D. H.: Nature (London) **179** (1957) 1252.
[2] Phaff, H. J.: In: The Yeast — Physiology and Biochemistry of Yeast (A. H. Rose and J. S. Harrison, Editors) Vol. 2, 135, Academic Press London and New York 1971.
[3] Villanueva, J. R. and Acha, J. G.: In: Methods in Microbiology (C. Booth, Editor) Vol. 4, 666, Academic Press London—New York 1971.
[4] Sugimori, T., Uchida, Y. and Tsukada, Y.: Agric. biol. Chem. **36** (1972) 669.
[5] Blechschmidt, D. und Tröger, R.: Z. Allg. Mikrobiol. **14** (1974) 245.
[6] Rademacher, J. H. und Reuter, G.: Z. Allg. Mikrobiol. **18** (1978) 107.
[7] Kitamura, K., Kaneko, T. and Yamamoto, Y.: Arch. Biochem. Biophysics **145** (1971) 402.
[8] Kitamura, K., Kaneko, T. and Yamamoto, Y.: J. gen. appl. Microbiol. **18** (1972) 57.
[9] Kitamura, K., Kaneko, T. and Yamamoto, Y.: J. gen. appl. Microbiol. **20** (1974) 323.
[10] Doi, K., Doi, A. and Fukui, T.: J. Biochem. (Tokyo) **70** (1971) 711.
[11] Doi, K., Doi, A. and Fukui, T.: Abric. biol. Chem. **37** (1973) 1619.
[12] Jones, G. H. and Ballou, C. E.: J. biol. Chemistry **244** (1969) 1043.
[13] Vrsanska, M., Kratky, Z. and Biely, P.: Z. Allg. Mikrobiol. **17** (1977) 391.
[14] Vrsanska, M., Kratky, Z. and Biely, P.: Z. Allg. Mikrobiol. **17** (1977) 465.
[15] Reuter, G., Hässler, O., Wisniewski, H. G. und Röber, B.: DDR-Patent WP C 12 D/215195 (1979).
[16] Paleg, L. G.: Anal. Chem. **31** (1959) 1902.
[17] Lowry, O. H., Rosebrough, N. J., Farr, A. L. and Randall, R. J.: J. biol. Chemistry **193** (1951) 265.
[18] Popov, B., Reuter, G. and Meyer, H. W.: Z. Allg. Mikrobiol. **20** (1980) 47.

Acta Biotechnologica 2 (1982) 1, 103—106

SHORT COMMUNICATIONS

Calorimetric Glucose Determination by Glucose Oxidase — Methylene Blue

D. Kirstein, L. Schilder and E. Kahrig

Akademie der Wissenschaften der DDR
Zentralinstitut für Molekularbiologie
Bereich Angewandte Enzymologie
1115 Berlin-Buch, Lindenberger Weg 70

Summary

The reaction between glucose and methylene blue, catalyzed by glucose oxidase (GOD), was analysed calorimetrically.
The amount of heat produced under saturating methylene blue concentrations ($> 10^{-2}$ mol/l) was measured with glucose concentration and time as parameters (kinetic procedure). Kinetic constants (pseudo one substrate kinetics) were derived from the experimental data: $K_{M(glucose)} = 1.18 \times 10^{-3}$ mol/l and $V_{max} = 0.085$ J/mg GOD min ($3.89 \cdot 10^{-6}$ mol/mg GOD min).
Comparison of caloric with optical measurements gave an enthalpy of reaction of 22.52 kJ/mol. Considering the observed substrate inhibition, glucose determinations are possible up to glucose concentrations of 0.1 mol/l.

Zusammenfassung

Die Glukoseoxydase-katalysierte Reaktion zwischen Glukose und Methylenblau wurde kalorimetrisch verfolgt.
Die unter sättigenden Methylenblaukonzentrationen ($> 10^{-2}$ mol/l) gebildeten Wärmemengen wurden mit Glukosekonzentration und Zeit als Parametern bestimmt (kinetisches Verfahren). Die Experimente ergaben folgende kinetische Daten (Pseudo-Einsubstratkinetik): $K_{M(Glukose)} = 1{,}18 \cdot 10^{-3}$ mol/l und $V_{max} = 0{,}085$ J/mg GOD · min ($3{,}89 \cdot 10^{-6}$ mol/mg GOD · min).
Der Vergleich der kalorischen mit optischen Messungen ergab eine Reaktionsenthalpie von 22,52 kJ/mol. Unter Berücksichtigung der beobachteten Substratüberschußhemmung sind Glukosebestimmungen bis zu 0,1 mol/l möglich.

Introduction

Glucose determinations in body fluids by enzymatic procedures have been described in several papers (e. g. [1—5]).
The application of immobilized enzymes in enzyme thermistors [6, 7] or in enzyme reactors [3—5, 8—10] stimulated the development of caloric methods both for analytical aims and for determination of kinetic constants. The upper limit of the glucose oxidase catalyzed reaction between glucose and oxygen [3—5, 7] is determined by the low solubility of oxygen. The heat production of this process approaches a limit value which de-

pends on the initial oxygen concentration and on the presence or absence of catalase (H_2O_2-decomposition) [7].
The aim of our experiments was the extension of the range of measurable glucose concentrations to higher limits by application of a second substrate showing better solubility than oxygen. In this paper are presented the results obtained with soluble GOD.

Materials and methods

Glucose oxidase from *P. notatum* (VEB Arzneimittelwerk Dresden, activity: 11.75 U/mg, 25 °C), glucose DAB 7 (VEB Maisanwerk Barby) and methylene blue Extra B (VEB Feinchemie Sebnitz, GDR) ε_{665}: 56400 cm²/mol) were used.
The measurements were performed in a 20 ml calorimeter as described in [11]. The calorimeter was calibrated by electrical heating. The temperature/time curve was registered by a "Universal-Temperature-Measuring-Device" (Fa. KNAUER, Berlin-West) and a compensation-recorder (type MKEI/J, VEB Meßgerätewerk Magdeburg). A "Specord UV-VIS" spectrophotometer (VEB Carl Zeiss, Jena) was used for optical measurements.
Enzyme and substrates were dissolved in acetate buffer ($\mu = 0.1$, pH = 5.6). The solutions were prethermostated to 25 °C.

Results

At first the reaction rate of the GOD catalyzed reaction between glucose and methylene blue was examined. Whereas the reaction between oxygen and glucose is finished in 8 seconds under comparable conditions the methylene blue oxidation takes more than 1 hour. To use of values of temperature-time slopes (initial slopes) instead of reaction enthalpies is therefore advantageous for the glucose analytics both with respect to the stability of thermostating and for saving time.
Generally we used the increase of temperature within the first 10 minutes as basis for the calculation of the heat production by the reaction. Comparison with optical measurements showed that within 10 minutes nearly 42% of the glucose reacted with methylene blue.
The slope of the curve in fig. 1 is similar to a simple MICHAELIS-MENTEN-kinetics up to 0.25 mol/l glucose. At higher glucose concentrations the results show the influence of a glucose inhibition.
Figures with values of 1 and 5 minutes, respectively, have qualitatively the same trend, an influence of products is therefore less probable.
Variation of the methylene blue concentrations within $5 \cdot 10^{-3}$ and $5 \cdot 10^{-2}$ mol/l showed that practically there is substrate saturation for methylene blue in this range. Enzyme kinetic parameters can be determined from the values of fig. 1 by LINEWEAVER BURK plot between glucose concentrations of $5 \cdot 10^{-4}$ and $2.5 \cdot 10^{-2}$ mol/l:

$$K_{M(\mathrm{glucose})} = 1.18 \cdot 10^{-3}\ \mathrm{mol/l}\ \text{and}$$
$$V_{\max} = 0.0844\ \mathrm{J/mg\ GOD \cdot min} = 3.89 \cdot 10^{-6}\ \mathrm{mol/mg\ GOD \cdot min}$$

The MICHAELIS-MENTEN curve constructed with these constants is included in fig. 1 as a dashed line.
The molar enthalpy of the reaction was found by comparison of calorimetrically and optically determined reaction rates. The decrease of methylene blue absorption at 665 nm at a glucose concentration of 0.25 mol/l gave a reaction rate of $2.12 \cdot 10^{-3}$ mol/l min.
The corresponding caloric value is 9.55 J/min and therefore the enthalpy of reaction is 22.52 ± 0.18 kJ/mol.

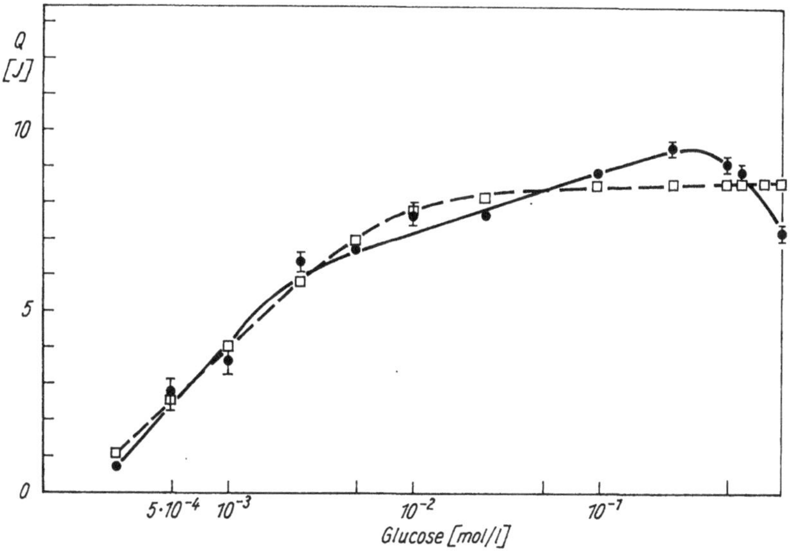

Fig. 1. Reaction enthalpy of glucose oxidase catalyzed glucose oxidation by methylene blue (within 10 min of reaction) as a function of glucose concentration in a semi-logarithmic plot

—○— : measured course

—□— : calculated course with

$K_{M(\mathrm{Glucose})} = 1.18 \cdot 10^{-3}$ mol/l

$V_{\max} = 0.085$ J/mol GOD min

$C_{\text{methylene blue}} = 0{,}05$ mol/l

$C_{\mathrm{GOD}} = 0.5$ mg/ml

Initial temperatur: 25°C

Discussion

The enthalpy of reaction of the GOD catalyzed glucose oxidation by methylene blue is about one quarter of the enthalpy of the reaction when oxygen is used to convert glucose to gluconolacton enzymatically (83.74 kJ/mol [4]). Nevertheless the method described here is useful for analytical purposes because glucose concentrations up to 0.1 mol/l can be measured directly (oxygen: $7 \cdot 10^{-5}$ mol/l [4]), due to the better solubility of the second substrate. Ambiguity of the results due to substrate excess inhibition can be eliminated if the glucose concentration is nearly known.

The kinetic data show that the maximal reaction rate for methylene blue is 4 times smaller (3.89 U/mg GOD) than for oxygen (11.75 U/mg GOD).

The apparent K_M-value of glucose with methylene blue is nearly one order of magnitude smaller than with oxygen as the second substrate.

Eingegangen: 5. 3. 1981

References

[1] GOLDBERG, R. N., PROSEN, E. J., STAPLES, B. R., BOYD, R. N., ARMSTRONG, G. T., BERGER, R. L., YOUNG, D. S.: Anal. Biochem. **64** (1975) 68.
[2] MCGLOTHLIN, C. D., JORDAN, J.: Anal. Chem. **47** (1975) 786.
[3] JOHANSSON, A.: Protides Biol. Fluids **20** (1973) 567.
[4] SCHMIDT, H. L., KRISAM, G., GRENNER, G.: Biochim. Biophys. Acta **429** (1976) 283.
[5] KRISAM, G.: Fresenius Z. Anal. Chem. **290** (1978) 130.
[6] MOSBACH, K., DANIELSSON, B.: Biochim. Biophys. Acta **364** (1974) 140.
[7] TRAN-MINH, C., VALLIN, D.: Anal. Chem. **50** (1978) 1874.
[8] REHAK, N. N., EVERSE, J., KAPLAN, N. O., BERGER, R. L.: Anal. Biochem. **70** (1976) 381.
[9] CANNING, L. M., CARR, P. W.: Anal. Lett. **8** (1975) 359.
[10] FERRESTE, L. J.: Anal. Lett. **7** (1974) 599.
[11] BARK, L. S., BARK, S. M.: Thermometric Titration. Pergamon Press, N. Y. 1969.

Acta Biotechnologica 2 (1982) 1, 107–113

SHORT COMMUNICATIONS

Einfluß von Penicillin und seinen Zersetzungsprodukten auf die Biosynthese von L-Lysin mit Brevibacterium flavum CB

M. Bučko, V. Revallová, A. Hano, A. Cunderlíková, M. Miklåš

BIOTIKA — Slovenska Lupca, CSSR

Paper given on 2nd Symposium of Socialist Countries about Biotechnology, Leipzig 2.—5. 12. 80

Summary

In the laboratory scale fermentation the substitution of peanutmeal by a hydrolysate of *Penicillium* mycelium in the culture medium for production of L-Lysine with *Brevibacterium flavum* CB has been testet. The mycelium hydrolysate contains Penicillin and degradation products of Penicillin; therefor the influence of these substances to production of L-Lysine has been investigated.

Zusammenfassung

Unter Laborfermentationsbedingungen wurde untersucht, inwieweit bei der L-Lysin-Biosynthese mit dem Produktionsstamm *Brevibacterium flavum* CB im Nährmedium Erdnussmehl-Hydrolysat durch ein Hydrolysat von *Penicillium*-Mycel ersetzt werden kann. Da in diesem Hydrolysat Penicillin und Zersetzungsprodukte des Penicillins enthalten sind, wurde deren Einfluß auf die L-Lysin-Produktion getestet.

Einleitung

In der letzten Zeit steigt in der Welt der Verbrauch von Aminosäuren besonders für die Tierernährung und in der Human- und Veterinärmedizin. Die industrielle Gewinnung von Aminosäuren bedient sich hauptsächlich biosynthetischer und chemisch-synthetischer Verfahren.

Im volkseigenen Betrieb Biotika, Slovenská Lupca wird L-Lysin biosynthetisch mit dem Produktionsstamm *Brevibacterium flavum* CB [1] hergestellt. An stickstoffhaltigen Substraten werden Maisextrakt und Erdnussmehl-Hydrolysat eingesetzt. Als weiterer stickstoffhaltiger Nährstoff kommt das Hydrolysat des Mycels des Produktionsstammes *Penicillium chrysogenum* in Betracht, welches außer Proteinen Vitamine, Enzyme und weitere biologisch aktive Stoffe enthält. Letztgenanntes Hydrolysat enthält weiterhin Penicillin und Zersetzungsprodukte des Penicillins. Aus diesen Gründen befaßten wir uns mit dem Einfluß des Penicillins und seiner Zersetzungsprodukte auf die Biosynthese von L-Lysin.

Penicillin und einige weitere Antibiotika sind dadurch bekannt, daß sie in die Synthese der Peptidglykane der bakteriellen Zellwand eingreifen und zwar in verschiedenen

Stadien dieser Synthese [2]. Darum wirken die Penicilline nur auf wachsende bakterielle Zellen. Nichtwachsende Zellen werden nicht angegriffen [3, 4].
Bei der Biosynthese der Aminosäuren beeinflußt Penicillin die Permeabilität der Zelloberfläche, welche auch für die Ausschüttung der Aminosäuren bedeutsam ist. Die Zugabe des Penicillins in die Nährlösung zu Beginn der Fermentation inhibiert die Bildung des L-Lysins etwa in gleichem Maße wie bei der Kultivierung des Produktionsstammes auf einem Biotin nicht enthaltenden Medium. Mit Hilfe des Penicillins ist es also möglich, den Verlauf der Fermentation zu regulieren [5].

Einfluß von Penicillin auf die Biosynthese von L-Lysin

a) Zugabe verschiedener Mengen von Benzylpenicillin in den Produktionsnährboden zu Beginn der Laborfermentation (Abb. 1):

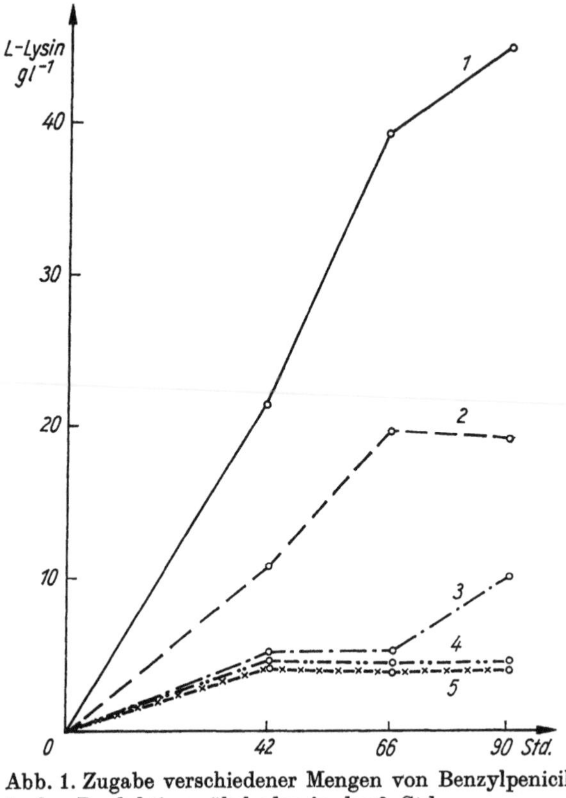

Abb. 1. Zugabe verschiedener Mengen von Benzylpenicillin in den Produktionsnährboden in der 0. Std.

Benzylpenicillin I. E. ml⁻¹		Biomasse g l⁻¹		
		42. Std.	66. Std.	90. Std.
1	0	21,3	39,6	45,3
2	0,5	10,4	19,7	19,0
3	1	4,6	4,8	9,9
4	2	4,3	3,6	3,5
5	5	4,2	3,5	3,3

Die Zugabe von 0,5 I. E. Benzylpenicillin auf 1 ml des Produktionsnährbodens zu Beginn der Fermentation verzögert das Wachstum; die Konzentration von 5 I. E. ml^{-1} bringt das Wachstum des Produzenten fast zum Erliegen.

b) Zugabe 1 I. E. Benzylpenicillin auf 1 ml des Produktionsnährbodens in verschiedenen Stunden der Kultivierung (Laborfermentation); (Abb. 2):

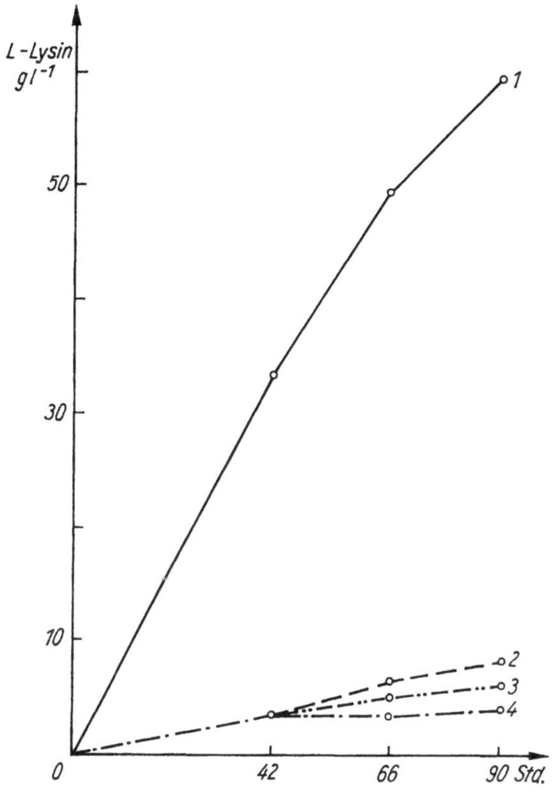

Abb. 2. Zugabe 1 I. E. Benzylpenicillin auf 1 ml des Produktionsnährbodens in verschiedenen Stunden der Kultivation

Benzylpenicillin		L-Lysin		Glutaminsäure
I. E. ml^{-1}		Std.	g l^{-1}	g l^{-1}
1	0	0.	59,9	1,2
2	1	0.	8,5	5,9
3	1.	4.	4,7	9,7
4	1	8.	6,7	20,7

Die Zugabe von 1 I. E. Benzylpenicillin auf 1 ml des Produktionsnährbodens in der 0., 4. und 8. Stunde inhibiert die Produktion des L-Lysins und verursacht eine Ausscheidung von Glutaminsäure. Die höchste Konzentration an Glutaminsäure wurde bei Zugabe Benzylpenicillins in der 8. Std. gefunden.

c) Zugabe verschiedener Mengen von Benzylpenicillin in den Produktionsnährboden in der 30. Std. (20-l-Laborfermentor) (Abb. 3):

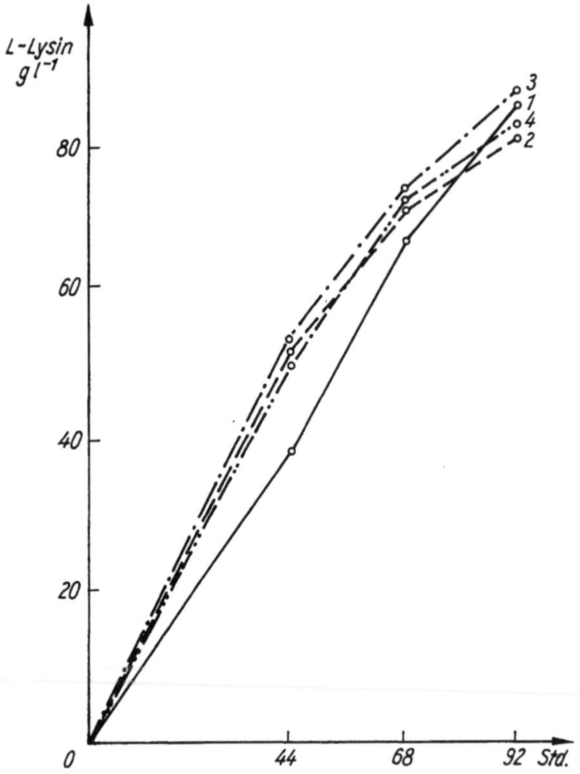

Abb. 3. Zugabe verschiedener Mengen von Benzylpenicillin in den Produktionsnährboden in der 30. Std.

Benzylpenicillin		L-Lysin ($g\ l^{-1}$)			
I. E. ml^{-1}		Std.	44. Std.	68. Std.	92. Std.
1	0	0.	48,0	66,3	85,3
2	5	30.	52,2	70,8	82,1
3	10	30.	53,2	73,2	86,0
4	20	30.	52,0	71,7	82,7

Benzylpenicillin wirkt nicht inhibierend, wenn es in die Kulturlösung nach Beendigung der Wachstumsphase zugegeben wird.

d) Einfluß der im sauren Milieu entstandenen Zersetzungsprodukte des Benzylpenicillins auf die Produktion des L-Lysins (Laborfermentation) (Abb. 4):

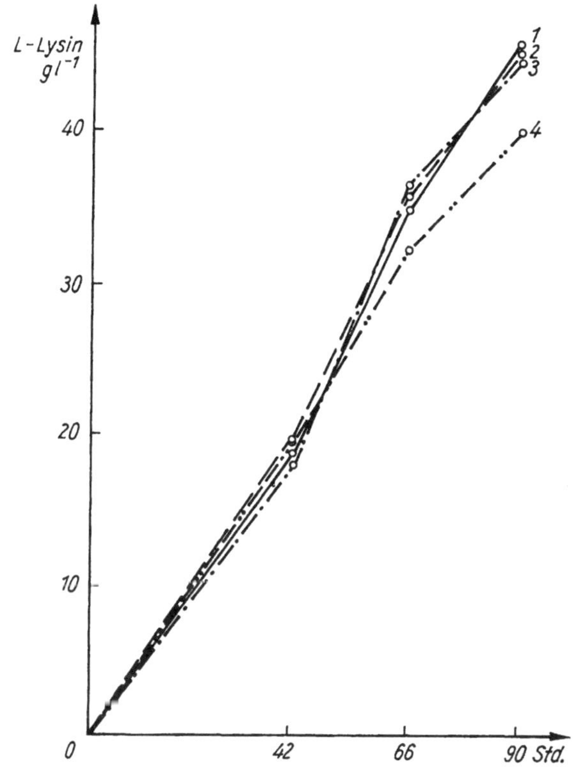

Abb. 4. Einfluß der im sauren Milieu entstandenen Zersetzungsprodukte des Benzylpenicillins auf die Produktion des L-Lysins

Zersetzungsprodukt von x I. E.		L-Lysin $g\,l^{-1}$		
Benzylpenicillin		42. Std.	66. Stunde	90. Std.
1	0	19,0	35,1	45,5
2	30	19,0	34,2	4,9
3	90	18,6	35,8	45,2
4	155	19,6	32,5	40,4

Erst bei Zugabe der aus 155 I. E. Benzylpenicillin im sauren Milieu entstandenen Zersetzungsprodukte pro ml Produktionsnährmedium kommt es zu einer mäßigen Inhibition der Produktion des L-Lysins.

e) **Einfluß der im alkalischen Milieu entstandenen Zersetzungsprodukte des Benzylpenicillins auf die Produktion des L-Lysins (Laborfermentation) (Abb. 5):**

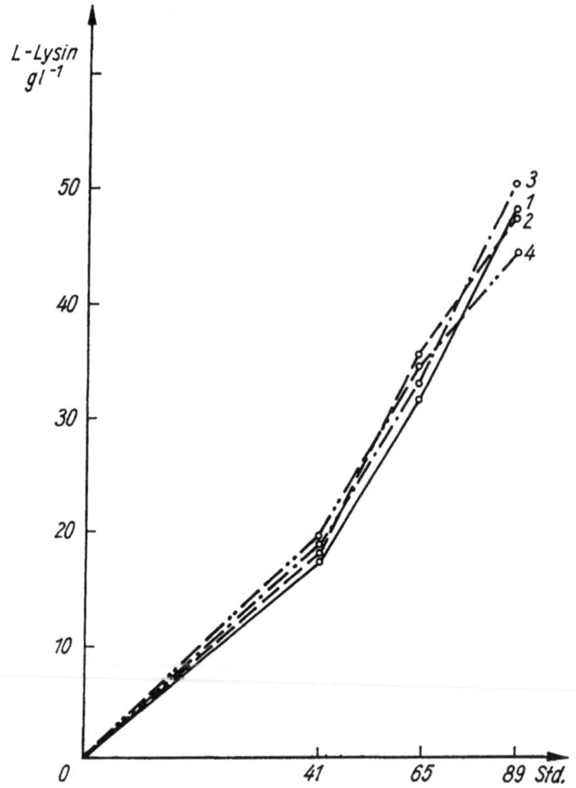

Abb. 5. Einfluß der im alkalischen Milieu entstandenen Zersetzungsprodukte des Benzylpenicillins auf die Produktion des L-Lysins

Zersetzungsprodukt von x I. E. Benzylpenicillin		L-Lysin g l^{-1}		
	x	42. Std.	66. Std.	90. Std.
1	0	17,4	32,7	49,5
2	30	17,4	35,6	49,4
3	90	17,8	33,9	50,8
4	155	17,9	35,2	45,6

Enthält der Fermentationsnährboden in 1 ml Zersetzungsprodukte, die im alkalischen Milieu aus 155 I. E. Benzylpenicillin entstanden sind, kommt es in den letzten Stunden der Fermentation zu einer mäßigen Inhibition der L-Lysin-Produktion.

f) Verwendung einer Mischung aus Erdnußmehl-Hydrolysat und Penicillium-Mycel-Hydrolysat für die Zubereitung des Produktionsnährbodens zur Biosynthese von L-Lysin (Fermentation im 20-l-Laborfermentor):

	Erdnußmehl-Hydrolysat 1/10 l Nährboden	Penicillium-Mycel-Hydrolysat 1/10 l Nährboden	L-Lysin g l^{-1} 90. Std.
1	3,14	0	79,7
2	1,94	1,94	81,0
3	1,74	1,74	84,3
4	2,55	0,85	83,2

Bei der Zubereitung des Nährbodens kann man einen Teil des Erdnußmehl-Hydrolysats durch Penicillium-Mycel-Hydrolysat ersetzen.

Schlußfolgerungen:

Es wurde bestätigt, daß das zu Beginn der Fermentation und während der Wachstumsphase zugegebene Penicillin eine Senkung der Produktion des L-Lysins und eine Ausschüttung von Glutaminsäure bewirkt.
Bei Zugabe von Zersetzungsprodukten (pro ml Fermentationsmedium), die im sauren oder alkalischen Milieu aus 90 I. E. Benzylpenicillin entstanden sind, kommt es noch nicht zu einer Inhibierung der L-Lysin-Produktion.
Es wurde nachgewiesen, daß für die Zubereitung des Nährbodens zur Biosynthese von L-Lysin das saure Hydrolysat eines *Penicillium*-Mycels teilweise das teure Erdnußmehl-Hydrolysat ersetzen kann.

Literatur

[1] Bučko, M., Beňuška, N.: Biologizace a chemizace vyzivy zvířat **12** (1976) 2, 141.
[2] Krčmery, V. u. a.: Antibiotiká, Bratislava, 1978, 26.
[3] Strominger, J., Izaki, L., Matsuhashi, K.: Fermentation Proceedings **26** (1967) 9.
[4] Strominger, J.: Ann. New York Acad. Sci. **235** (1974) 210.
[5] Yamada, K., Kinoshita, S., Tsunoda, T.: The Microbial Production of Aminoacids; Kodansha LTD, Tokyo 1972, 40.

PATENT REPORTS

DD-WP 145279 (C12N-1/02)
AN: Akademie der Wissenschaften der DDR; DD
ER: Teich, W.; Skrowny, H.; DD
AD: 19. 9. 79/BD: 3. 12. 80

Verfahren zur Abtrennung von Mikroorganismen aus Suspensionen

Ist applizierbar zur Gewinnung von Mikroorganismenbiomasse und umfaßt die kombinierte Anwendung spezifischer, physiologisch unbedenklicher Koagulationshilfsmittel (Alkalisilikatlösungen, vorzugsweise Natriumwasserglas, allein bzw. zusammen mit Ammoniak und Kalziumsalzen) und thermischer Behandlung bei vorzugsweise 70–150 °C und einem Druck von vorzugsweise 0,98–5 bar. Die koagulierte Mikroorganismensuspension wird mit verdünnter Mineralsäure, besonders Phosphor- und/oder Schwefelsäure oder Salzsäure, behandelt und ein pH-Wert von 4–7 eingestellt.
5 Ausführungsbeispiele (zur Gewinnung von Bakterienkonzentrat); 11 Seiten

DD-WP 146616 (C12N-1/00)
AN: Akademie der Wissenschaften der DDR; DD
ER: Hadeball, W.; Rühlemann, I.; Pörschmann, S.; Scheibe, P.; Heinritz, H.-J.; Ringpfeil, M.; DD
AD: 6. 7. 79/BD: 18. 2. 81

Verfahren zur Gewinnung von Biomasse aus Gülle

Bezieht sich besonders auf Hühner- und Schweinegülle, bei gleichzeitiger Reduzierung unerwünschter Inhaltsstoffe, durch Kultivierung von Reinkulturen (z. B. *Candida utilis*), definierten Mischkulturen sowie Mischpopulationen, in denen eine Art (z. B. *Pseudomonas* sp.) mit 60–80% (nach der Zellzahl) dominiert. Die Gülle wird einer einstufigen kontinuierlichen Fermentation, gegebenenfalls unter Zusatz von externen C- und P-Quellen unterworfen. Durch die Zusätze wird das C : N : P-Verhältnis in der Gülle so gestaltet, daß es einem Verhältnis von 25 : 5 : 1 nahekommt, so daß eine maximale mikrobielle Biomassesynthese bei gleichzeitigem maximalen Abbau organischer und anorganischer Inhaltsstoffe erreicht wird.
Die applizierten Mikroorganismen werden bei pH von 3–8, einer Temperatur von 20–60 °C, besonders zwischen 30 und 40 °C und einer Verweilzeit von vorzugsweise zwischen 3 und 7 Stunden kultiviert. Die gewonnene Biomasse zeichnet sich durch hohen Rohprotein- und relativ niedrigen Aschegehalt aus.
7 Ausführungsbeispiele; 16 Seiten

PATENT REPORTS

DE-OS 2844398 (B01f-3/04)
AN: Uhde GmbH; Heinrich Frings GmbH & Co. Kg; DE
ER: Ebner, H.; Faust, U.; Sittig, W.; DE
AD: 12. 10. 78/OD: 22. 5. 80

Verfahren und Vorrichtung zum Dispergieren eines Gases in einer Flüssigkeit

Flüssigkeit durchströmt in vertikal geführtem Kreislauf zwei in einem Reaktor parallel angeordnete Reaktionsräume. Vorzugsweise zur Belüftung bei aeroben mikrobiologischen Prozessen, insbesondere bei Verfahren zur Herstellung von SCP geeignet, beinhaltet unter Ausschaltung negativer Faktoren bekannter Verfahren und Vorrichtungen, daß ein Teilstrom der absinkenden Flüssigkeitsmenge aus dem Unterteil des zweiten Reaktionsraumes (I) durch eine am Reaktorboden angeordnete Dispergiervorrichtung (II) mit durch diese angesaugtem Gas vermischt und das Gemisch mittels eines den Rotor der (II) umgebenden Leitkanalkranzes in den unteren Teil des ersten (I) ausgeschleudert wird, und daß dieses ausgeschleuderte Gemisch beim Aufwärtsströmen mit dem Hauptstrom der im zweiten in den ersten (III) übertretenden Flüssigkeitsmenge zusammengeführt wird.
2 Ausführungsbeispiele (Applikation bei methanolutilisierendem *Methylomonas clara*); 2 Zeichnungen; 20 Seiten

DE-OS 2854557 (B01F-3/04)
AN: Institut techniceskoj teplofiziki Akademii Nauk Ukrainskoj SSR; SI
ER: Kremnev, O. A.; Dolinskij, A. A.; Korcinskij, A. A.; u. a.; SU
AD: 18. 12. 78/OD: 18. 9. 80

Verfahren zur Belüftung von Flüssigkeit und Vorrichtung (Bioaerator mit im wesentlichen senkrecht angeordnetem Gaszuführungsstutzen (I))

Verfahren unter Druckpulsationsbedingungen, bei denen Betriebs- und Bauart ein großes Druckgefälle in der Flüssigkeit zur Zerstörung des Films von oberflächenaktiven Stoffen bei den meisten Gasblasen gewährleisten und gleichzeitig die Bildung einer stabilen Gasflüssigkeitsemulsion im (I) ohne Zuführung schaumbildender Reagentien sichern. Umfaßt die portionsweise Gaszu- bzw. -abführung aus dem (I), wobei sich die Phasengrenze im (I) mit einer Frequenz von 40—200 Schwingungen/Min. und einer Amplitude von 0,05—2 m bewegt. Verfahren und Vorrichtung sind in der mikrobiologischen Industrie, z. B. bei der Fermentation von Futterantibiotika, applizierbar.
5 Ausführungsbeispiele; 13 Zeichnungen; 44 Seiten

PATENT REPORTS

DE-OS 2924181 (C12M-1/06)
AN: Intermedicat GmbH; CH
ER: FIECHTER, A.; GERLACH, K.; DE
AD 15. 3. 79/OD: 8. 1. 81

Fermentor zum Züchten von Mikroorganismen

Gefäß, in das von unten her ein Rührwerk hineinragt und in dessen oberem Bereich sich ein rotierender Schaumabschneider (I) befindet, der mit einer Gasauslaßleitung in Verbindung steht, mit einem koaxial angeordneten, oben und unten offenen Leitzylinder, einem Begasungsrohr mit mehreren Stechanschlüssen in der Gefäßwand für Zugabe und Entnahme von Gasen oder Flüssigkeiten. Durch die Anordnung (dicht unter der oberen Wand des Gefäßes mit vertikalem Abstand über dem Leitzylinder) des (I) und mittels fächerförmiger Lamellen zum Zerschlagen des Schaumes ist der Fermentor, ohne daß permanent mit Schaum gefüllte Zonen entstehen, praktisch vollständig mit einem Gemisch aus Nährflüssigkeit und Gas in gleichmäßiger Verteilung zu füllen.
1 Ausführungsbeispiel; 7 Zeichnungen; 24 Seiten

DE-OS 2921918 (C12N-1/38)
AN: KLÖCKNER-HUMBOLDT-DEUTZ AG; DE
ER: ALBERT, A.; DE
AD: 30. 5. 79/OD 4. 12. 80

Verfahren zur Optimierung der Stoffwechselaktivität von Mikroorganismen im Substrat eines biologischen Reaktionssystems

Insbesondere bei der Faulung von Schlämmen einer biologischen Abwässerkläranlage, wobei ein Teil des im Durchlauf durch das Reaktionssystem (I) fermentierten Substrates abgezweigt und vorzugsweise unter Mischung mit frischem Substrat (II) in das (I) rezirkuliert wird. Beinhaltet die mindestens teilweise Entwässerung des abgezweigten Substrates mit Hilfe eines künstlichen Schwerefeldes, vorzugsweise unmittelbar vor der Mischung mit frischem (II). Man unterwirft (II) im Schwerefeld einer Beschleunigung von insbesondere zwischen 450 und 650 g. Die Einwirkungszeit des Schwerefeldes beträgt vorzugsweise weniger als 60 Sek., um eine Beeinträchtigung der Stoffwechselbedingungen der Mikroorganismen zu vermeiden. Vorteile der Erfindung sind z. B. die Optimierung der Stoffwechselvorgänge sowohl in der sauren als auch in der alkalischen Fermentation; Schaffung der Voraussetzungen für eine vollautomatische Steuerung und damit für die Prozeßautomatisierung.
1 Ausführungsbeispiel; 2 Zeichnungen; 24 Seiten

Correction from Acta Biotechnologica 1 (1981) 4, 391:

2. Symposium der sozialistischen Länder über Biotechnologie vom 2.—5. Dezember 1980 in Leipzig. Das 3. Symposium der sozialistischen Länder über Biotechnologie wird vorraussichtlich im Frühjahr 1983 in der ČSSR durchgeführt.

Contents

Short Communication

Bei Fragen zur Produktsicherheit wenden Sie sich bitte an:
If you have any questions regarding product safety,
please contact:

Walter de Gruyter GmbH
Genthiner Straße 13
10785 Berlin
productsafety@degruyterbrill.com